CE Conformity Marking

and New Approach Directives

To Clive and Kay, owners of the only Prostar with CE Marking!

About the Author

Ray Tricker (MSc, IEng, FIE(elec), FInstM, MIQA, MIRSE) as well as being the Principal Consultant and Managing Director of Herne European Consultancy Ltd – a company specialising in ISO 9000 and ISO 14000 Management Systems – is also an established Butterworth-Heinemann author. He served with the Royal Corps of Signals (for a total of 37 years) during which time he held various managerial posts culminating in being appointed as the Chief Engineer of NATO ACE COMSEC.

Most of Ray's work since joining Herne has centred on the European Railways. He has held a number of posts with the Union International des Chemins de fer (UIC) (e.g. Quality Manager of the European Train Control System (ETCS), European Union (EU) T500 Review Team Leader, European Rail Traffic Management System (ERTMS) Users Group Project Co-ordinator, HEROE Project Co-ordinator) and is currently preparing a complete Quality Management System for the European Rail Research Institute (ERRI) in Holland, aimed at gaining them ISO 9000 accreditation in the near future. He is also consultant to the Association of American Railroads (AAR) advising them on ISO 9001:12000 compliance.

CE Conformity Marking

and New Approach Directives

Ray Tricker

OXFORD AUCKLAND BOSTON JOHANNESBURG MELBOURNE NEW DELHI

Butterworth-Heinemann
Linacre House, Jordan Hill, Oxford OX2 8DP
225 Wildwood Avenue, Woburn, MA 01801-2041
A division of Reed Educational and Professional Publishing Ltd

A member of the Reed Elsevier plc group

First published 2000

British Library Cataloguing in Publication Data
A catalogue record for this book is available from the British Library

Library of Congress Cataloguing in Publication Data
A catalogue record for this book is available from the Library of Congress

ISBN 0 7506 4813 9

Composition by Genesis Typsetting, Rochester, Kent
Printed and bound in Great Britain by Biddles Ltd, www.biddles.co.uk

Contents

Preface

Up until 1985, the European Community had always tried to remove technical barriers by attempting to harmonise technical product manufacturing specifications as opposed to setting performance levels. Unfortunately, this meant that highly technical specific instruments were required for each product category. This approach became even more difficult to develop and control, as it required endless technical debates. That is why the Commission submitted to the Council a 'new approach' to technical harmonisation and standards.

On 7 May 1985 the Council of European Communities adopted a Resolution concerning the harmonisation of European Directives and standards. This Resolution (called 'New Approach') was aimed at the technical harmonisation of European Directives and the removal of Europe's internal barriers to trade and the free movement of goods (i.e. customs duties, taxes, quantitative restrictions and measures, national monopolies, state aid and tax discrimination). During December 1989 this Resolution was expanded to a broader, more global, approach to conformity assessment. The following year (in Directive 90/683/EEC) the Council reconfirmed their commitment and laid down that all industrial products covered by New Approach Directives may only be placed on the market after the manufacturer has affixed a conformity mark to them.

On 22 July 1993, the Council published the current CE Conformity Marking Directive under 93/465/EEC, which states that:

> The aim of the CE Marking is to symbolise the conformity of a product with the levels of protection of collective interests imposed by the total harmonisation Directives and to indicate that the economic operator has undergone all the evaluation procedures laid down by Community law in respect of his product.

All New Approach (and many of the Old Approach) Harmonised Directives are affected by this requirement.

The prime aim of *CE Conformity Marking and new approach directives* is to enable readers to understand the requirements of the CE Directive and to see how it affects the European Directives. This book also provides guidance on the sort of Quality Management System that a manufacturer and/or supplier will need in order to be able to work in conformance with these requirements and can be used to pave the way for companies wanting to make the transition, cost effectively, from CE Marking to ISO 9001:2000.

The main parts of the book are as follows:

- **Background to New Approach Directives**
  - A potted history of European Harmonised Directives and standards, their production, management and distribution plus their inter-relationship with International standards.
  - An overview of the application of New Approach Directives and their extension to include other European and world-market countries through Mutual Recognition Agreements and European Conformity Assessment Protocols.
  - The effect of CE Marking and Conformity Assessment.
- **Structure of New Approach Directives**
  - A full description of the principles and main elements that make up New Approach Directives.
- **Structure of the CE Conformity Marking Directive**
  - Translating legal 'Eurospeak' into everyday language together with an explanation of the main requirements of the Directive.
  - A full description of the Conformity Assessment Modules (and their variants) that are associated with CE Marking.
  - The essential requirements for CE Marking (e.g. its form, how it should be displayed, affixed and withdrawn).
  - Responsibilities of Competent Bodies and Notified Bodies.
- **The requirements of the various Directives affected by CE Marking**
  - A description of the format of New Approach Directives and an overview of the principal Directives including their structure, objectives, essential requirements, proof of conformity and their individual requirements for CE Marking and Quality Control.
- **Gaining CE Conformity**
  - A full description of the Conformity Assessment Procedures and their associated requirements for a documented Quality Management System.
  - Details of the manufacturers' tasks under each conformity module and the basic requirements for manufacturers of industrial products.
- **Glossary**
  - A list of the most frequently used terms and their definitions.

- **References**
  - A listing of all the main documents and standards utilised in this publication including their availability and function.
- **Acronyms and Abbreviations**
  - It must be said that the CE Conformity Marking Directive is not one of the easiest documents to read(!) as it tends to use a lot of 'Eurospeak' terminology and acronyms and abbreviations (see Figure i). For ease of reference, therefore, a list of the main acronyms and abbreviations has been included.
- **Addresses**

  Lists showing:
  - Where to obtain published standards
  - National and international standards
  - Information sources

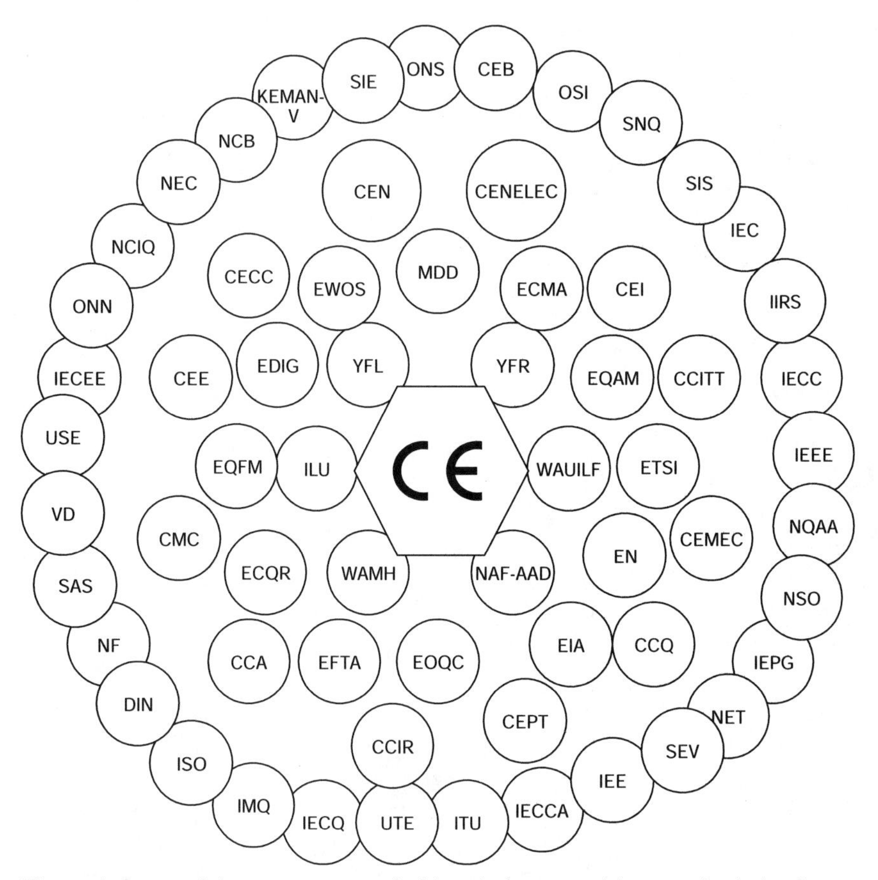

Figure i Some of the acronyms and abbreviations used by standards bodies

For convenience (and in order to reduce the number of equivalent or similar terms) the following are considered interchangeable terms within this book:

- product/appliance/machine/equipment;
- manufacturer/supplier;
- notified body/competent body/third party;
- product/device.

1 BACKGROUND TO NEW APPROACH DIRECTIVES

Although harmonisation and European standardisation commenced in 1957 with the Treaty of Rome (and its aim of progressively establishing the internal market over a period expiring on 31 December 1992), it wasn't until 1985 that the European Economic Community (EEC) – now the European Union (EU) – initiated the 'New Approach for Harmonised Directives and Standards'.

Prior to 1985, however, the Community had always tried to remove technical barriers by trying to harmonise technical product manufacturing specifications as opposed to setting performance levels. Unfortunately, this meant that highly technical, system-specific-standards were required for each product category. Over the years, this approach became even more difficult to develop and control, as it required endless technical debates as trade, quality and safety differed amongst almost all of the EEC countries. Over the years, many Community members had also established their own laws and standards and even when the EEC tried to harmonise these standards within Europe, national deviations still existed. That is why the Commission submitted to the Council a 'New Approach' to technical harmonisation and standardisation.

Figure 1.1 New Approach Directives

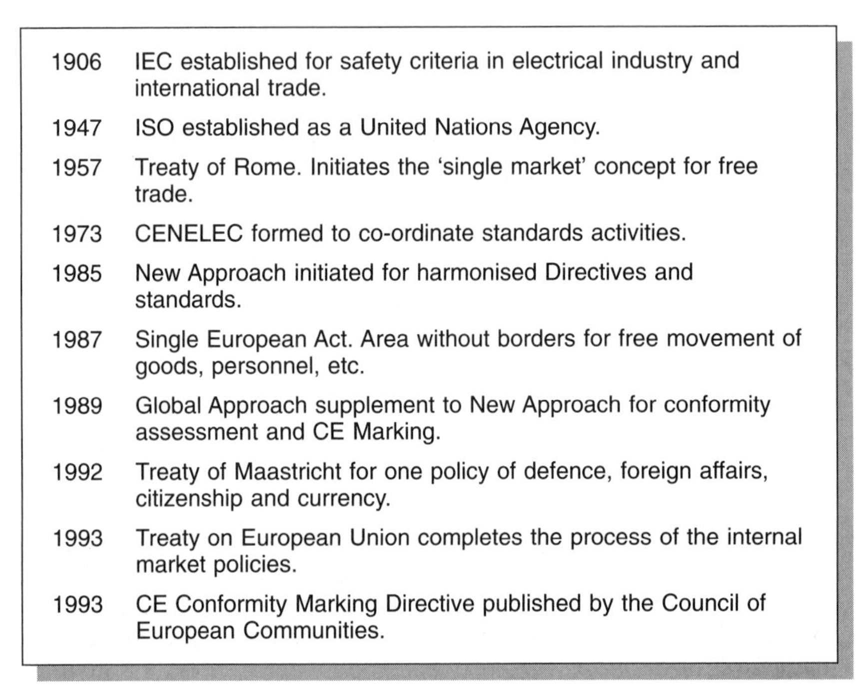

1906	IEC established for safety criteria in electrical industry and international trade.
1947	ISO established as a United Nations Agency.
1957	Treaty of Rome. Initiates the 'single market' concept for free trade.
1973	CENELEC formed to co-ordinate standards activities.
1985	New Approach initiated for harmonised Directives and standards.
1987	Single European Act. Area without borders for free movement of goods, personnel, etc.
1989	Global Approach supplement to New Approach for conformity assessment and CE Marking.
1992	Treaty of Maastricht for one policy of defence, foreign affairs, citizenship and currency.
1993	Treaty on European Union completes the process of the internal market policies.
1993	CE Conformity Marking Directive published by the Council of European Communities.

Table 1.1 The history of European harmonisation

The Council approved the policy in its resolution of 7 May 1985 and agreed that the internal market would 'comprise an area without internal frontiers in which the free movement of goods, persons, services and capital (i.e. customs duties, taxes, quantitative restrictions/measures, national monopolies, state aid and tax discrimination etc.) was ensured and that restrictions on imports would be prohibited between Member States'.

With the introduction of the New Approach Directives came the initial goal for the total harmonisation of all regulations throughout the Community.

New Approach Directives became the methods by which the Council of the European Commission (EC) can ensure that the products that have been legally manufactured or marketed in one country can be moved freely throughout the Community. The prime aim of these Directives is to remove barriers to trade whilst allowing minor differences that occur between national requirements (e.g. legislation) to still be acceptable – provided that they:

- serve a legitimate purpose (particularly with regard to health, safety, consumer protection, environmental protection); and
- can be justified.

In December 1889 this Resolution was expanded to provide a more flexible approach to conformity assessment. The following year the Council reconfirmed their commitment and laid down that **all** industrial products covered by New Approach Directives may be placed on the market **only** after the manufacturer has affixed a conformity mark to them. Initially this conformity mark was referred to as the 'EC Mark'. With the gradual introduction of the New Approach Directives, however, this was then changed to the 'CE Mark' but, because of religious connotations, the term has now been finalised as '**CE Marking**'.

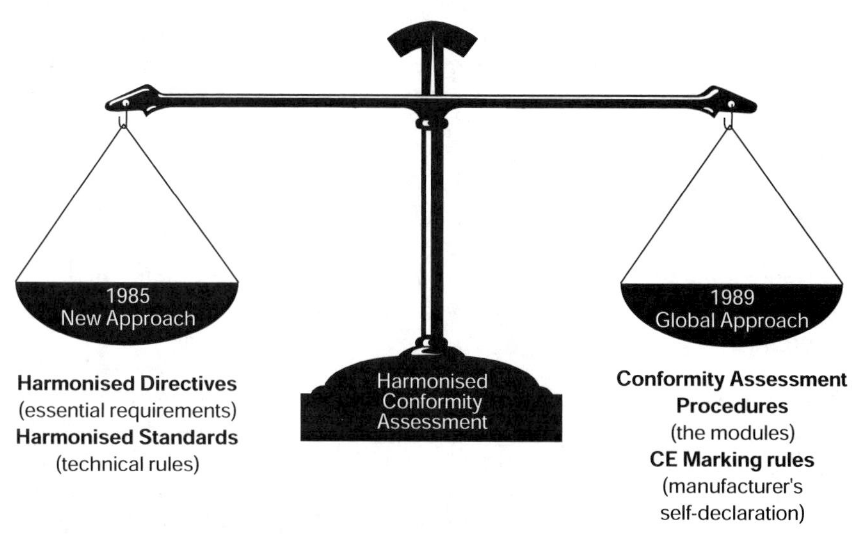

Figure 1.2 The New and Global Approaches

The overall objective of the New Approach Directives is to provide flexible conformity assessment procedures (see 1.17) over the entire manufacturing process that can be adapted to the needs of an individual operation. As a supplement to this objective, Council decision 90/683/EEC introduced a more modular, 'global' approach, which subdivided conformity assessment into a number of operations (i.e. modules). The modules contained in this Directive differ according to the stage of development of the product (e.g. design, prototype, full production), the type of assessment involved (e.g. documentary checks, type approval, quality assurance), and the person carrying out the assessment (i.e. the manufacturer or a third party).

The Global Approach was then replaced and brought up to date by decision 93/465/EEC (i.e. the CE Marking Directive) which laid down general guidelines and detailed procedures for conformity assessment that are to be used in New Approach Directives based on:

- a manufacturer's internal design and production control activities;
- third party type examination combined with a manufacturer's internal production control activities;
- third party type (or design) examination combined with:
  - third party approval of a product;
  - production quality assurance systems;
  - third party product verification.
- third party unit verification of design and production;
- third party approval of full quality assurance systems.

One of the prime requirements of the 1989 Global Approach is that **all** industrial products must conform to the Essential Requirements (ERs) of the relevant Directives before being placed on the market. The transposition of Directives and harmonised standards into national laws and standards then makes them legally binding and obligatory within the Member States. The ERs are:

- lay down the necessary elements for protecting public interest;
- are mandatory and only products complying with the Essential Requirements may be placed on the market and/or put into service;
- must be applied as a function of the risks (hazards) inherent with a given product.

The New Approach has not been applied to sectors where Community legislation was well advanced prior to 1985, or where provisions for finished products and risks related to such products cannot be laid down. (For instance, Community legislation on foodstuffs, chemical products, pharmaceutical products, motor vehicles and tractors do not follow the principles of New Approach.) New Approach Directives are based on the following principles:

- harmonisation is limited to Essential Requirements;
- only products fulfilling the Essential Requirements are subject to free movement;
- harmonised standards, the reference numbers of which have been published in the *Official Journal of the European Communities* (*OJEC*) and which have been transposed into national standards, are presumed to conform to the corresponding Essential Requirements;
- application of harmonised standards or other technical specifications remains voluntary, and manufacturers are free to choose any technical solution that provides compliance with the Essential Requirements;
- manufacturers may choose between different conformity assessment procedures provided for in the applicable Directive;
- products manufactured in compliance with harmonised standards shall be recognised as being in conformity with the corresponding Essential Requirements.

The harmonised standards associated with the New Approach are designed to offer a guaranteed level of protection for the Essential Requirements listed in the Directives. National Authorities are then responsible for ensuring that the safety and/or other interests listed in the Directive are covered. A procedure (i.e. a safeguard clause) allows Member States to challenge the conformity of a product in addition to failures or shortcomings of harmonised standards.

Most of the New Approach Directives incorporate a 'Self-Certification' option. This leaves the responsibility for applying the test standards, and certifying that the products comply, with the product manufacturer. Whilst this means that there is, in these instances, no mandatory requirement for third party intervention, in practice, many manufacturers will contract part of the assessment process to a third party laboratory.

Originally, a third party assessment was considered necessary only where products were not manufactured in compliance with harmonised standards. Since the first New Approach Directives were adopted, however, this has now mainly been overcome.

1.1 European Directives

European Directives are the laws that manufacturers must meet before they are permitted to affix CE Marking to their products. The Directives are identified with the year and identifier number such as '89/336/EEC' for ElectroMagnetic Compatibilty (EMC), which means that it was the 336th Directive to be published in 1989. Similar to other Directives, it was then subjected to a short transitional period (normally this is 3 or 4 years) before finally being adopted. So although the EMC Directive was published on 3 May 1989, it was not adopted and brought fully into force by the Member States until the end of 1995.

European Directives are the policy decisions made by the Council of European Communities and tell us why we must comply and what could (will probably!) happen if we choose to ignore the laws. They describe the Essential Health and Safety Requirements that suppliers must meet before equipment is placed on the market and what must be done from a procedural and legal point of view.

The principles of complying with the requirements of New Approach are similar for all Directives. Namely:

- products should comply with the applicable 'Essential Requirements'. This is usually a matter of product design, and product instructions;
- the manufacturer or supplier should demonstrate that the product complies with the Essential Requirements, using one of the product conformity assessment modules available within the Directive. One of the options is usually the assessment of the product by applying the

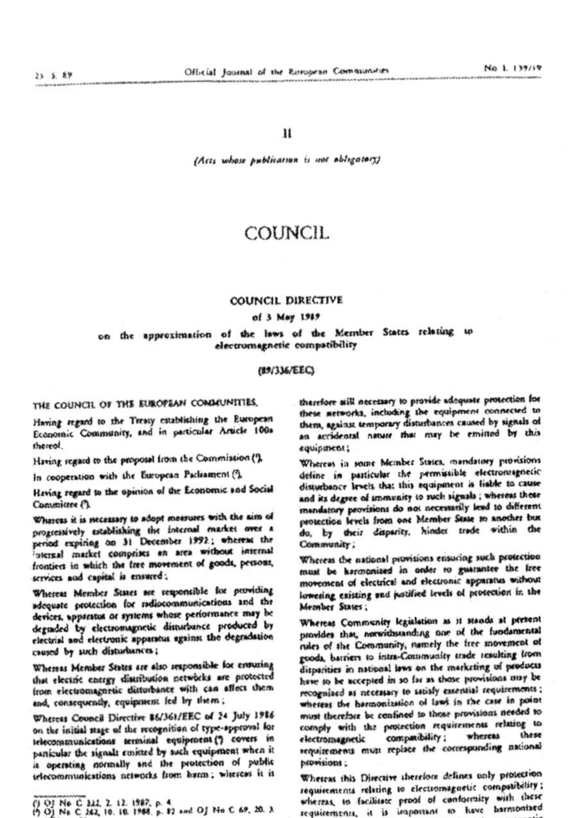

25. 5. 89 Official Journal of the European Communities No L 139/19

II

(Acts whose publication is not obligatory)

COUNCIL

COUNCIL DIRECTIVE

of 3 May 1989

on the approximation of the laws of the Member States relating to electromagnetic compatibility

(89/336/EEC)

THE COUNCIL OF THE EUROPEAN COMMUNITIES,

Having regard to the Treaty establishing the European Economic Community, and in particular Article 100a thereof,

Having regard to the proposal from the Commission ([1]),

In cooperation with the European Parliament ([2]),

Having regard to the opinion of the Economic and Social Committee ([3]),

Whereas it is necessary to adopt measures with the aim of progressively establishing the internal market over a period expiring on 31 December 1992; whereas the internal market comprises an area without internal frontiers in which the free movement of goods, persons, services and capital is ensured;

Whereas Member States are responsible for providing adequate protection for radiocommunications and the devices, apparatus or systems whose performance may be degraded by electromagnetic disturbance produced by electrical and electronic apparatus against the degradation caused by such disturbances;

Whereas Member States are also responsible for ensuring that electric energy distribution networks are protected from electromagnetic disturbance with can affect them and, consequently, equipment fed by them;

Whereas Council Directive 86/361/EEC of 24 July 1986 on the initial stage of the recognition of type-approval for telecommunications terminal equipment ([4]) covers in particular the signals emitted by such equipment when it is operating normally and the protection of public telecommunications networks from harm; whereas it is therefore still necessary to provide adequate protection for these networks, including the equipment connected to them, against temporary disturbances caused by signals of an accidental nature that may be emitted by this equipment;

Whereas in some Member States, mandatory provisions define in particular the permissible electromagnetic disturbance levels that this equipment is liable to cause and its degree of immunity to such signals; whereas these mandatory provisions do not necessarily lead to different protection levels from one Member State to another but do, by their disparity, hinder trade within the Community;

Whereas the national provisions ensuring such protection must be harmonized in order to guarantee the free movement of electrical and electronic apparatus without lowering existing and justified levels of protection in the Member States;

Whereas Community legislation as it stands at present provides that, notwithstanding one of the fundamental rules of the Community, namely the free movement of goods, barriers to intra-Community trade resulting from disparities in national laws on the marketing of products have to be accepted in so far as those provisions may be recognized as necessary to satisfy essential requirements; whereas the harmonization of laws in the case in point must therefore be confined to those provisions needed to comply with the protection requirements relating to electromagnetic compatibility; whereas these requirements must replace the corresponding national provisions;

Whereas this Directive therefore defines only protection requirements relating to electromagnetic compatibility; whereas, to facilitate proof of conformity with these requirements, it is important to have harmonized standards at European level concerning electromagnetic compatibility, so that products complying with them may

([1]) OJ No C 322, 2. 12. 1987, p. 4.
([2]) OJ No C 262, 10. 10. 1988, p. 82 and OJ No C 69, 20. 3. 1989, p. 72.
([3]) OJ No C 134, 24. 5. 1988, p. 2.
([4]) OJ No L 217, 5. 8. 1986, p. 21.

Figure 1.3 Example Directive showing publication date and identifier number

relevant standards to the product, and/or the preparation of appropriate documentation such as the Technical Construction File;
- CE Marking should be affixed legibly and indelibly to the product;
- the manufacturer should prepare and sign a Declaration of Conformance.

The Directives:

- provide a high level of safety for a given product or product sector;
- set the range of possible choices;
- cover such issues as:
  - the appropriateness of the modules to the type of products;
  - the nature of the risks involved;
  - the economic infrastructures of the given sector (e.g. existence or non-existence of third parties);
  - the types (and importance of) production, etc.;

- set out the criteria governing the conditions for manufacture (leaving the manufacturer to choose the most appropriate modules for his protection);
- avoid imposing unnecessary requirements, which would be too onerous (however, be careful, this doesn't always happen!);
- enable the manufacturer to have a wide choice for ensuring compliance with the requirements of the various applicable Directives.

Many of the Directives include provision for the appointment of Notified Bodies. These are organisations appointed by the Member States in which they are based and whose details are 'Notified' to the European Commission. The Commission must then publish these details in the *OJEC* and Notified Bodies will perform specific functions (as defined by the Directives) in relation to the assessment of compliance of specific products. For many of the Directives their involvement is only mandatory for higher-risk and safety-critical products.

1.2 Transitional Period and Overlapping of Directives

EU product Directives are approved by the Member States, who then must transpose them into national law. The Directives define a schedule for adopting and publishing national provisions to implement each Directive. Directives also define when national provisions must be applied. A Directive is authorised when it has been published in the *OJEC*.

CE Marking Directives recognise a transitional period during which existing national provisions and new legislation will co-exist. In such cases, the manufacturer may choose to follow either the national or the new legislation. However, only the latter will allow the manufacturer to affix the CE Marking.

Some products may be governed by more than one Directive because different risks may be dealt with under different Directives. In cases where more than one Directive may apply, the CE Marking can be affixed only if the product complies with the appropriate provisions of **all** applicable Directives.

1.3 Application

The New Harmonised Directives apply to the countries shown in Figure 1.4.

1.3.1 EU Member States

All members of the Community have agreed to conform to the requirements of the New Approach Directives.

Figure 1.4 EU Member States

1.3.2 EFTA Countries

Three of the Member States of the European Free Trade Association (i.e. Iceland, Liechtenstein and Norway) have agreed to the requirements of the New Approach Directives. Switzerland did not want to be fully implicated but has now decided to seek 'Mutual Recognition Agreements' similar to the United States of America (see Section 1.3.4).

Figure 1.5 EFTA Countries

The EFTA Secretariat assists the Member States in activities under the Stockholm Convention (the legal basis for EFTA), in the day-to-day management of the Agreement on the European Economic Area (EEA) and the co-ordination and development of trade agreements with third countries.

EFTA States, and their economic operators, are subject to the same rights and obligations as their counterparts in the Community. For instance, the New Approach Directives are applied exactly in the same way in the EFTA States as in the Member States – although the administrative procedures concerning notification bodies and the safeguard clause are modified. Therefore, all guidance applicable to the Member States according to this guide also applies to the EFTA States.

1.3.3 Eastern European affiliates

Harmonisation and free movement of goods is also being discussed with the following countries (also see 1.4.2):

- Bulgaria;
- Cyprus;
- Czech Republic;
- Estonia;
- Hungary;
- Latvia;
- Lithuania;
- Poland;
- Romania;
- Slovakia;
- Slovenia;
- Turkey.

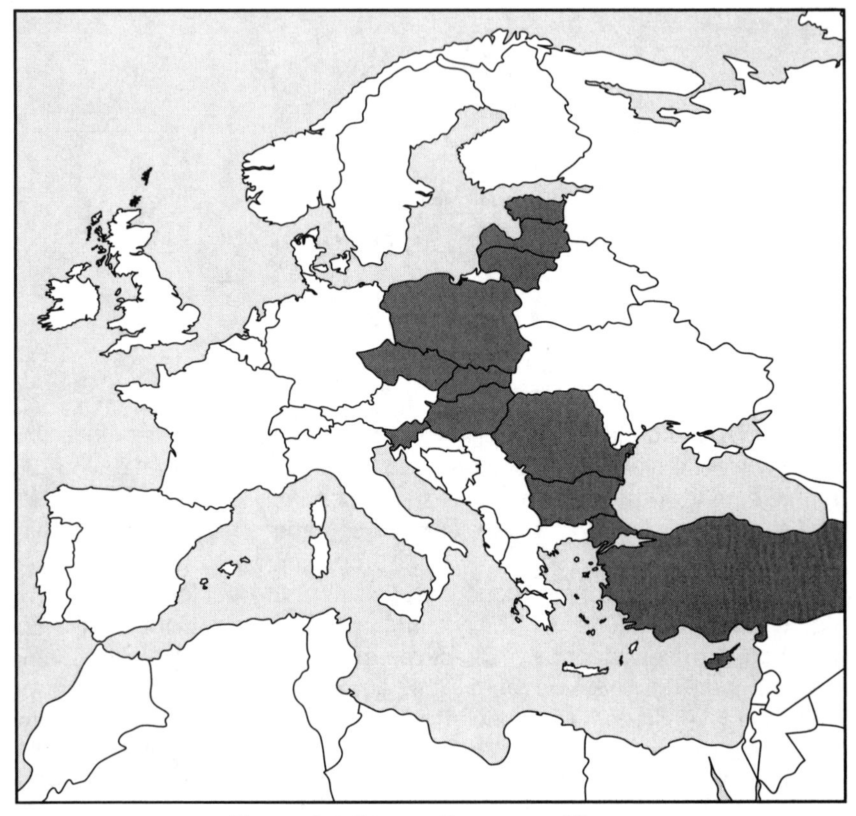

Figure 1.6 Eastern European affiliates

1.3.4 Others

Through 'Mutual Recognition Agreements' (see 1.4.1), the Community's market is being extended to:

- United States;
- Japan;
- Canada;
- Australia;
- New Zealand;
- Hong Kong;
- Israel;
- Singapore;
- Philippines;
- Republic of Korea;
- Switzerland (who opted not to sign the EEA (European Economic Area) agreement between the EU and EFTA – see 1.3.2).

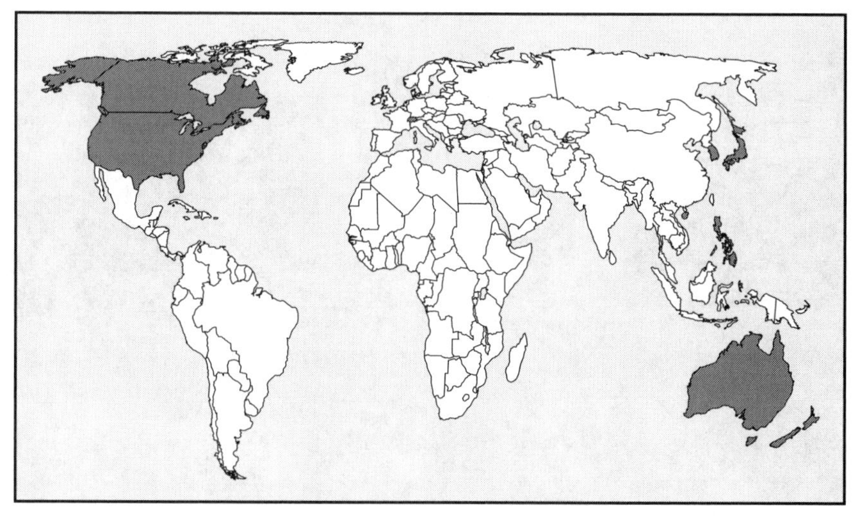

Figure 1.7 Other affiliates

1.4 Mutual Recognition Agreements/European Conformity Assessment Protocols

The Council's decision to extend the objective of New Approach Directives to include other European and world-market countries has resulted in two new programmes: namely Mutual Recognition Agreements and European Conformity Assessment Protocols.

1.4.1 Mutual Recognition Agreements

Mutual Recognition Agreements (MRAs) are aimed at extending the Community's market to other major industrial countries. They are established between the Community and the government of third countries that have a comparable level of technical development and have a similar approach to conformity assessment. They can apply to one or more categories of products or sectors covered by New Approach or other Community technical harmonisation Directives in force and, in certain cases, by non-harmonised national law.

Similar to New Approach Directives, MRAs comprise a framework agreement laying down the essential principles of the agreement supported by a number of annexes. These annexes specify the scope and coverage, regulatory requirements, the list of designated conformity assessment bodies, the procedures and authorities responsible for designating these bodies and, possibly, a transitional period.

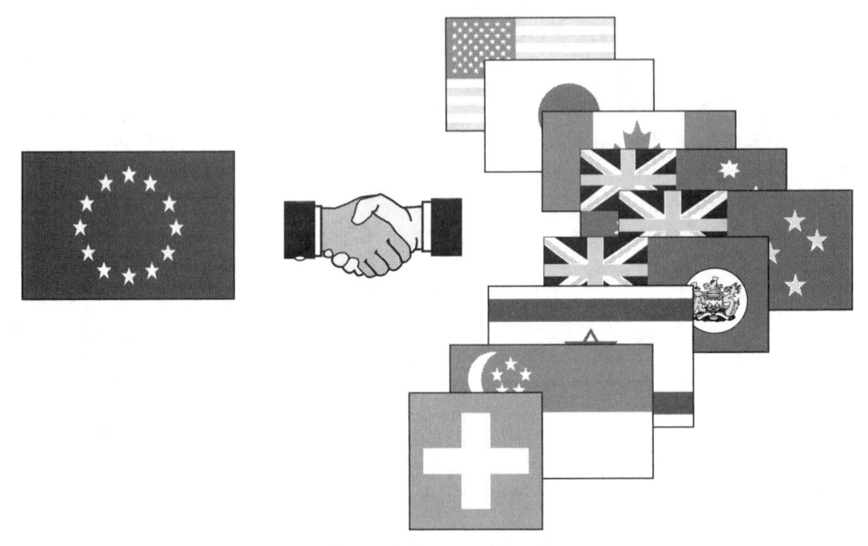

Figure 1.8 Mutual Recognition Agreements

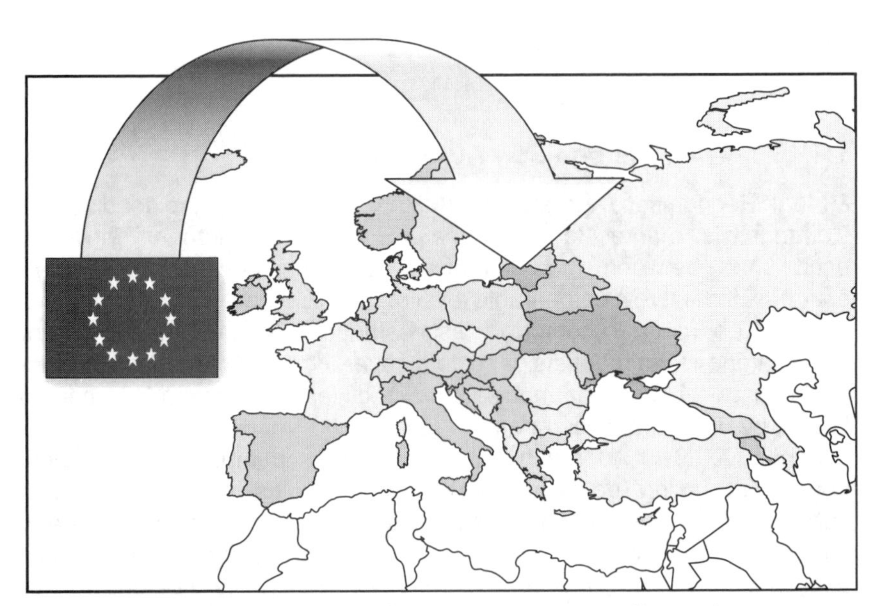

Figure 1.9 European Conformity Assessment Protocols

Currently the Commission has either finalised or is in the process of finalising agreements with, the United States, Japan, Canada, Australia, New Zealand, Hong Kong, Israel, Singapore, Philippines, Republic of Korea and Switzerland.

1.4.2 European Conformity Assessment Protocols

European Conformity Assessment Protocols (ECAP) are mutual recognition programmes that grant special status to Central and Eastern European countries who have either signed an association agreement with the Community (committing them to align their legislation with that of the Community) or who have applied for EU membership.

ECAPs comprise a framework agreement supported by a number of sectorial annexes and are seen as a means of promoting trade, health and safety between the EU and these countries as well as being a progressive extension of the Single Market. As part of the system, the Commission also provides the applicant countries with technical assistance programmes aimed at aligning their legislation with Community legislation (i.e. New Approach Directives and other harmonised technical legislation).

The countries currently concerned in this programme are Hungary, Poland, Czech Republic, Slovenia, Estonia, Cyprus, Romania, Bulgaria, Slovakia, Latvia and Lithuania.

1.5 Types of Directives and Standards

Directives, like standards, come in three forms: types A, B and C, otherwise called basic, generic, and product-specific. The type C product-specific standards are the top-level standards and take precedence over type A and B standards.

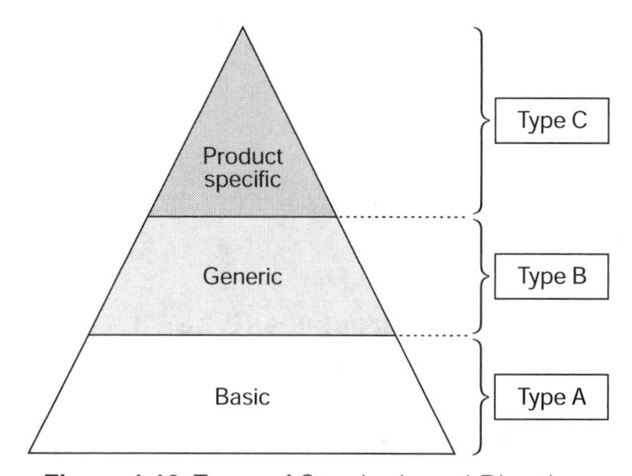

Figure 1.10 Types of Standards and Directives

A standard is considered harmonised at the time of announcement, which is when it is published in the *OJEC*. Compliance with the harmonised standards will, in most cases, ensure a product's conformity with the Essential Requirements of the Directives.

1.5.1 Basic Directives (type A)

Some of the basic Directives that apply to most product and machine suppliers are:

- General Product Safety Directive – 92/59/EEC;
- Product Liability Directive – 85/374/EEC;
- Conformity Assessment Procedures and CE Marking Rules – 93/465/EEC.

Basic standards associated with these Directives contain general principles for safe design or measurement techniques and levels, and may be applied to products when appropriate.

1.5.2 Generic Directives (type B)

These address a specific range or group of products, such as:

- Low-Voltage Directive (73/23/EEC);
- EMC Directive (89/336/EEC).

For machinery the B standards are further divided into B1 and B2 standards.

1.5.2.1 B1 Standards

These apply to particular aspects, such as surface temperatures and safety distances.

1.5.2.2 B2 Standards

These apply to particular safety devices or components such as Safety Switching Devices.

1.5.3 Product-Specific Directives (type C)

Apply to 'regulated' products, such as:

- Safety of Toys (88/378/EEC);
- Machinery (89/332/EEC);

- Medical Devices (93/42/EEC);
- Pressure Equipment (97/23/EC);
- Telecommunication Terminal Equipment (98/392/EEC).

1.6 Concurrent Application of Directives

New Approach Directives cover a wide range of products and/or risks, which both overlap and complement each other. As a result several Directives may have to be taken into consideration for one product (e.g. the Directive relating to machinery covers, in particular, mechanical hazards, the Directive relating to low voltage, electrical hazards, and the Directive relating to electromagnetic compatibility, electromagnetic disturbance and immunity).

In many circumstances, certain Directives may also make a direct reference to other Directives (e.g. the Directive concerning telecommunications terminal equipment is related to the Directive about low voltage equipment). In these cases, the application of a Directive may have to exclude certain Essential Requirements (or conformity procedures) of other Directives. This will usually require a risk analysis to be carried out on the product, or perhaps an analysis of the intended purpose of the product to determine the 'principle' Directive (e.g. the Directive relating to low voltage equipment is not applicable to electrical equipment for medical purposes; instead either the Directive relating to active implantable medical devices or medical devices may apply). In these circumstances, the placing on the market (and/or putting into service) can only take place when the product complies with the provisions of **all** of the applicable Directives and when the conformity assessment has been carried out in accordance with **all** those applicable Directives.

1.7 Aim of Standardisation

The aim of standardisation is **not** simply to produce paperwork that becomes part of a library! The aim of standardisation is to produce a precise, succinct, readily applied and widely recognised set of principles, which are relevant and satisfy the varied needs of business, industry and/or commerce.

Standardisation should also **not** give exclusive advantage to the products or services of one particular individual supplier and the application of standards should always be capable of objective verification by an independent third party evaluator (i.e. Certification Body).

But national bodies and national standards cannot dictate customer choice. A product that may legally be marketed need not be of universal appeal. Indeed, where different national standards persist they will do so as a reflection of different market preferences. For industry to survive in this new, 'liberalised' market, therefore, it must have a sound technological base supported by a comprehensive set of internationally approved standards.

Currently, the main producers of National Standards in Western Europe are:

- United Kingdom – British Standards Institution (BSI);
- Germany – Deutsch Institut fur Normung e.v. (DIN);
- France – Association Francais de Normalisation (AFNOR).

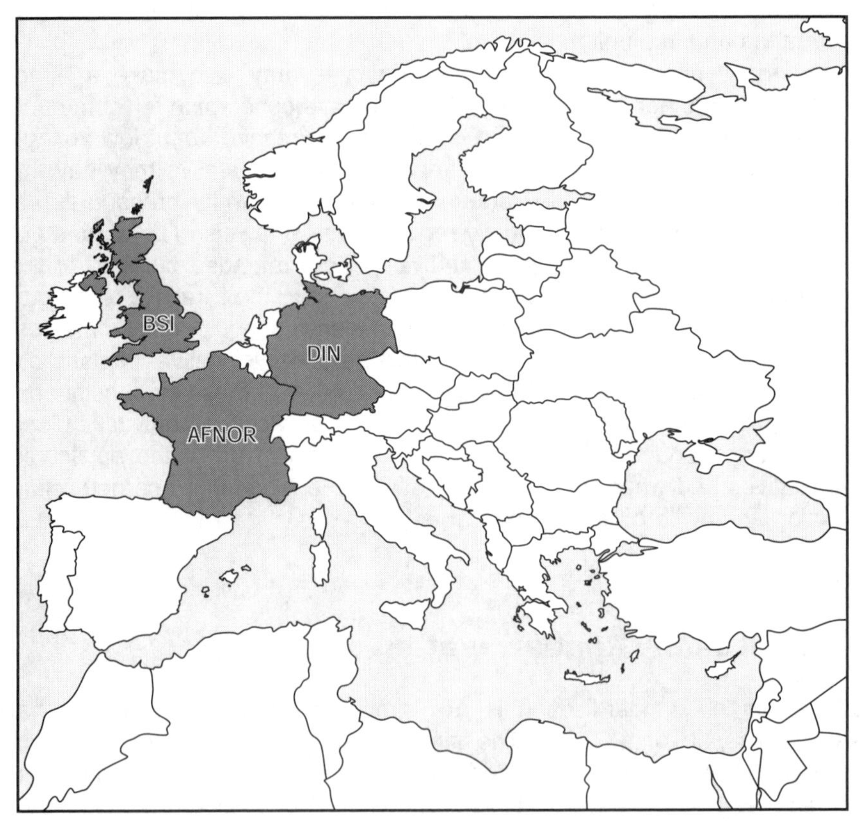

Figure 1.11 Main producers of national standards – Europe

Outside Europe the most widely used standards come from:

- America – American National Standards Institute (ANSI);
- Canada – Canadian Standards Association (CSA).

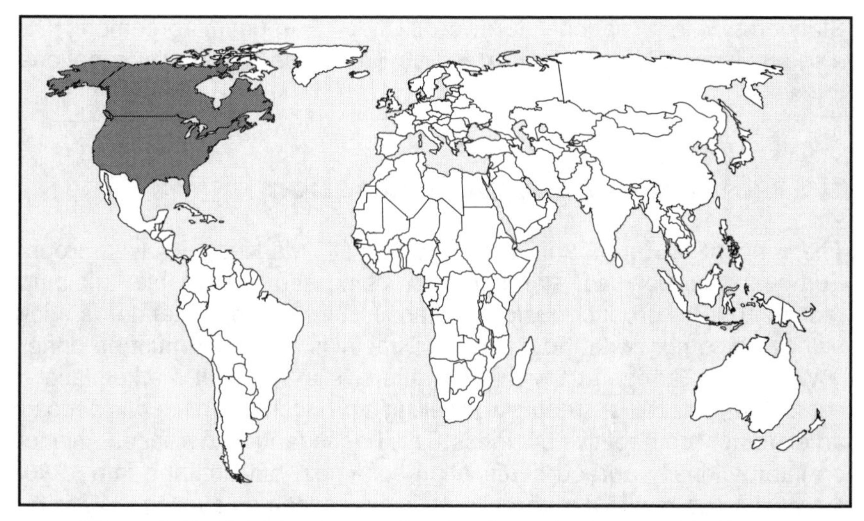

Figure 1.12 Main producers of national standards – outside Europe

Although these countries publish what are probably the most important series of standards, virtually every country with an industrial base has its own organisation producing its own set of standards.

CEN (Comité Europeén de Normalisation Electrotechnique – European Committee for Standardisation) and CENELEC (European Committee for Electrotechnical Standardisation – combination of CENEL and CENELCOM), together with ETSI (European Telecommunications Standards Institute) from the telecommunications field, are responsible for developing harmonised European standards that are crucial to the success of the European Single Market. In the broader international arena, it is the International Organisation for Standardisation (ISO) and the International Electrotechnical Commission (IEC) which pursue similar aims for harmonising world standards.

In all, there are more than 13,000 published ISO and IEC standards and, since the 1970s, the BSI has published most of these as British

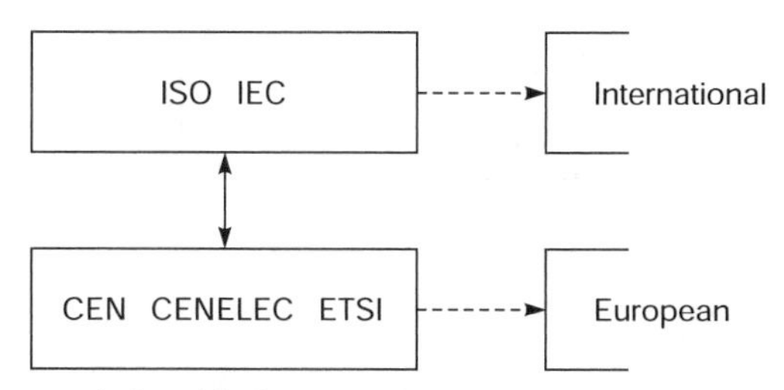

Figure 1.13 Inter-relationship between international and European standards

Standards with a national foreword. Under European agreement, BSI also publishes ENs as identical British Standards, again with a national foreword.

1.8 International Standardisation

The concept of European Conformity (i.e. CE Marking) revolves around European harmonised standards as being the acceptable minimum requirement for product design and assessment. These standards show us how to comply with the Council's Directives and what must be done.

Within the European Union (EU), there is a potential marketplace of some 350–400 million people and selling a product or service has become an extremely competitive business. This has meant an increased reliance on internationally agreed certification schemes, but until the late 1980s there were **no** viable third party certification schemes available. With the increased demand for assurance during all stages of the manufacturing processes, however, has come the requirement for manufacturers to work to recognised standards.

Over the years, therefore, there has been a steady growth in international standardisation and ISO (and the IEC) are now the standards bodies that most countries are affiliated to – through, that is, their own particular National Standards Organisation (NSO).

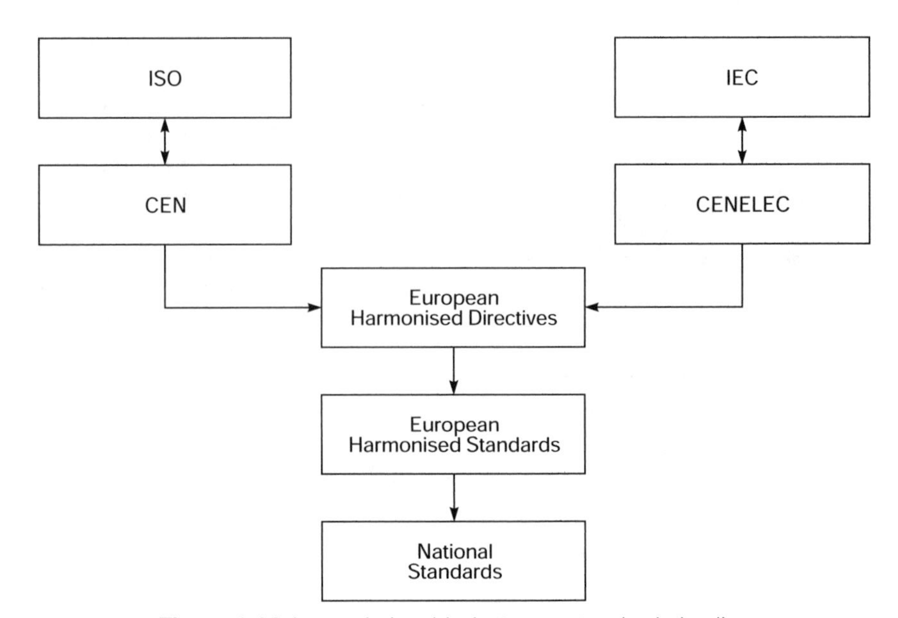

Figure 1.14 Inter-relationship between standards bodies

ISO (established as a United Nations Agency in 1947) is made up of representatives from more than 90 countries and includes the British Standards Institution (BSI) for the United Kingdom and ANSI (the American National Standards Institute for the United States). The work of ISO has increased considerably since it first got under way and a great number of standards are now available and have already been adopted.

These ISO and IEC standards (ISO is mainly concerned with industrial specifications and requests, whilst IEC refers to electrical equipment) were initially published as 'recommendations', but they are now accepted as international standards – in their own right – and the use of the word 'shall' (i.e. denoting a mandatory requirement) is becoming commonplace.

The international standards are, themselves, drawn up by International Technical Committees which have been approved by ISO or IEC member countries and there are now many hundreds of different ISO and IEC standards available, covering virtually every situation.

1.9 European Harmonised Standards

A harmonised standard is a technical specification (European standard, EN, or harmonisation document) that has been adopted by the European standards organisations (CEN, CENELEC and ETSI) and is prepared in accordance with the General Guidelines agreed between the Commission and the European standards organisations and following a mandate issued by the Commission.

Member States are required to transpose these harmonised standards, in full (i.e. the text of the European standard must be fully taken over) into

1. A mandate is drawn up, following consultation of the Member States.
2. The mandate is transmitted to European standards organisations.
3. European standards organisations accept the mandate.
4. European standards organisations elaborate a (joint) programme.
5. Technical Committee elaborates draft standard.
6. European standards organisations and national standards bodies organise public enquiry.
7. Technical Committee considers comments.
8. National standards bodies vote; European standards organisations ratify.
9. European standards organisations transmit references to the Commission.
10. Commission publishes the references.
11. National standards bodies transpose the European standard.
12. National authorities publish references of national standards.

Table 1.2 Adapter procedures

a national implementing standard, details of which have to be published in the *OJEC*. If there are any conflicting national standards in circulation, they must be withdrawn. However, it is acceptable to retain or publish a national standard dealing with a subject covered by the harmonisation document, **provided that** it has technically equivalent contents.

Harmonised documents and harmonised standards are not a specific category amongst European standards; the term only applies to standards associated with the New Approach Directives. They are a series of principles and commitments concerning standardisation, such as the participation of all interested parties (e.g. manufacturers, consumer associations, and trade unions), the role of public authorities, the quality of standards and a uniform application of standards throughout the Community.

Table 1.2 shows the adapter procedure for New Harmonised Directives.

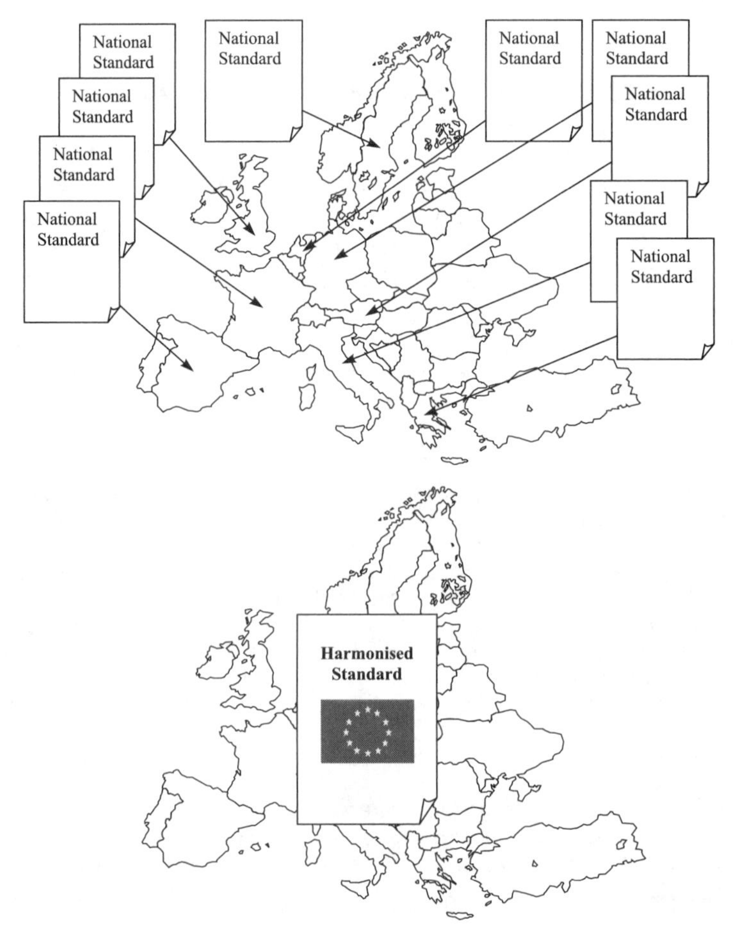

Figure 1.15 The benefits of harmonised standards for Europe

The difference between a 'harmonisation document' and a 'European standard' or 'EN' that provide a 'harmonisation standard' mainly concern the degree of obligation on the part of the Member States. Harmonisation documents must be implemented at national level, at least by public notification of the title and number of the document, and by the withdrawal of conflicting national standards. A product's conformity to harmonised standards means that a product is presumed to conform to the Essential Requirements of a New Approach Directive.

CEN, CENELEC and ETSI are considered competent bodies for adopting harmonised standards in accordance with the general guidelines defined by the Co-operation Agreement signed with the Commission of the European Communities on 13 November 1984.

1.10 Revision of Harmonised Directives

The formal decision to revise a standard is, in principle, taken by the European standards bodies. This takes place on the basis of their own initiative, or following a request from the Commission directly or, indirectly, based on an initiative of a Member State. The need for revision can result from:

- the changes of the scope of the Directive (e.g. an extension of the scope to other products or a modification of the Essential Requirements);
- the Commission or a Member State challenging the contents of the harmonised standard, indicating that it could no longer give presumption of conformity with the Essential Requirements; or
- technological development.

When a harmonised standard is revised, the revision must be covered by a mandate to maintain the possibility for giving a presumption for conformity. Unless the contrary can be deduced from the original mandate, the terms and conditions of the original mandate apply also for the revision of the harmonised standard. This does not exclude the possibility of a new mandate, in particular where the revision is related to shortcomings with respect to the Essential Requirements.

To give presumption of conformity, the revised standard must satisfy the general conditions according to the New Approach; i.e. the standard is based on a mandate, it is presented by the relevant European standards organisation to the Commission, its reference is published by the Commission in the *Official Journal*, and it is transposed as a national standard.

Following its internal regulations, the relevant European standards organisation lays down the date of publication at national level of the revised harmonised standard, and the date of withdrawal of the old standard. The transitional period is normally the time period between these two dates. During this transitional period both harmonised standards give presumption of conformity, provided that the conditions for this are met. After this transitional period, only the revised harmonised standard gives a presumption of conformity.

The Commission may consider that, for safety (and other reasons), the old version of the harmonised standard must cease giving a presumption of conformity before its withdrawal date (set by the European standards organisation in question) is past. In such cases, the Commission fixes an earlier date after which the standard will no longer give a presumption of conformity, and publishes this information in the *Official Journal.* If circumstances allow, the Commission consults the Member States prior to taking a decision to reduce or extend the period during which the standard gives a presumption of conformity.

The reference of the revised harmonised standard, together with the old harmonised standard and the date where the presumption of conformity of the old standard finishes, are published together in the *Official Journal.*

1.11 Management of the Lists of Standards

The three organisations responsible for providing the EC with harmonised standards are CEN, CENELEC, and ETSI;

1.11.1 CEN

CEN covers the same field as ISO, with its main sectors of activity being:

- information technology;
- biology and biotechnology;
- quality, certification and testing;
- transport and packaging;
- food;
- materials;
- chemistry;

Figure 1.16 CEN logo

- mechanical engineering;
- building and civil engineering;
- environment;
- health and safety at the workplace;
- gas and other energies;
- consumer goods, sports, leisure.

The CEN Certification Board also controls conformity assessment issues, in particular the CEN/CENELEC European Mark of conformity to European Standards (the Keymark).

Figure 1.17 Keymark logo

Currently CEN are developing certification schemes for the following four categories of product:

- plastic piping and ducting systems;
- heat cost allocators;
- ceramic floor and wall tiling;
- controllers for heating systems.

1.11.2 CENELEC

CENELEC is the European Committee for Electrotechnical Standardisation. It was set up in 1973 as a non-profit-making organisation under Belgian law and has been officially recognised as **the** European Standards Organisation in its field by the European Commission in Directive 83/189/EEC.

Figure 1.18 CENELEC logo

Its members have been working together in the interests of European harmonisation since the late 1950s and they work with 40,000 technical experts from 19 EU and EFTA countries to publish standards for the European market.

1.11.3 Technical Structure of CENELEC

All interested parties are consulted during the CENELEC standards drafting, through involvement in technical meetings at national and European level (to establish the content of the draft) and through enquiries conducted by the members.

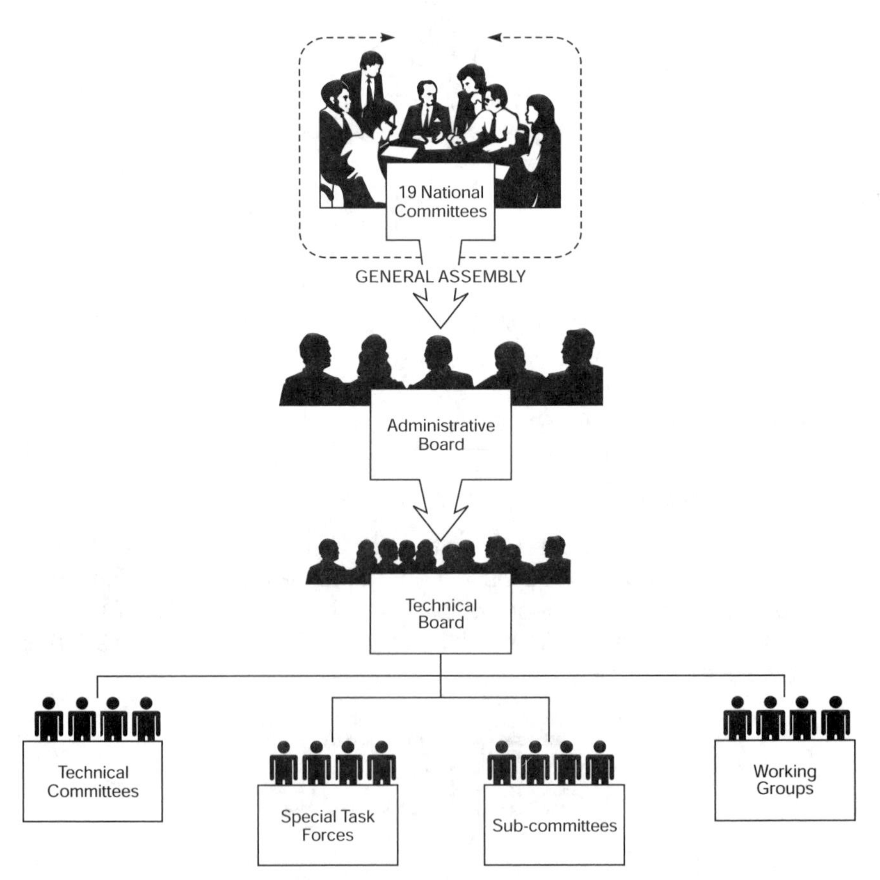

Figure 1.19 CENELEC – structure

1.11.3.1 General Assembly

The General Assembly (AG) is the highest-level body. It makes all the policy decisions and is composed of delegations from each of the 19 National Committees (NCs).

1.11.3.2 Administrative Board

An Administrative Board (CA) of eight officers, led by the President, supervises the work carried out according to the AGs resolutions.

1.11.3.3 Technical Board

The Technical Board (BT) co-ordinates the work of the technical bodies, which include Technical Committees (TCs), Sub-Committees (SCs), special Task Forces (BTTFs) and Working Groups (BTWGs). It is the BT, made up of one permanent delegate from each NC, which decides on ratification, on the basis of national voting, of draft standards prepared by the technical bodies.

The BT also approves work programmes and monitors the progress of standardisation work. The different CENELEC technical bodies are as follows:

1.11.3.4 Technical Committees

The Technical Committees (TCs) are established by the Technical Board with precise titles and scopes to prepare the standards. Technical Committees take into account any ISO/IEC work coming within their scope, together with such data as may be supplied by members and by other relevant international organisations, and work on related subjects in any other Technical Committees.

Each Technical Committee establishes and secures Technical Board approval for its programme of work with precise title, scope and scheduled target dates for the critical stages of each project. These dates are reviewed at least once a year.

1.11.3.5 Subcommittees

Subcommittees (SCs) may be established by a Technical Committee (after Technical Board approval on justification, programme of work, title and scope) having responsibility for a large programme of work in which:

- different expertise is needed for different parts of the work; and
- the range of separate activities needs co-ordination over long periods of time.

The parent TC retains full responsibility for the work of its SCs.

1.11.3.6 Technical Board Task Forces

The BTTFs (Technical Board Task Forces) are technical bodies set up by the Technical Board, with a view to undertake a specific short-term task within a target date and are composed of a Convenor and national delegations. A BTTF reports to the Technical Board, its parent body.

1.11.3.7 Technical Board Working Groups

The BTWGs (Technical Board Working Groups) are technical bodies set up by the Technical Board to undertake a specific short-term task within a target date. Its parent body disbands them once its task is completed. They are composed of a Convenor and of individual members appointed by the Technical Board and/or the National Committees to serve in a personal capacity.

1.11.3.8 Reporting Secretariats

Reporting Secretariats exist to provide information to the Technical Board on any ISO/IEC work which could be of concern to CENELEC. When the Technical Board wishes to examine a technical problem or to investigate a situation in an area not already covered by a Technical Committee, the Central Secretariat may initially call upon a Reporting Secretariat to provide what information is available. A Reporting Secretariat is undertaken by a CENELEC member, usually the member holding the Secretariat of the concerned IEC/TC or SC.

1.11.4 CENELEC Central Secretariat

Manned by over 40 people, the CENELEC Central Secretariat is a conglomerate of services designed to answer the needs for European standardisation and to serve the purpose of drafting, organising approval on and publishing European Standards. The present capacity of work volume exceeds more than one document ready for publication each calendar day.

1.11.5 ETSI

The European Telecommunications Standards Institute (ETSI) is a non-profit-making organisation whose mission is to determine and produce the telecommunications standards that will be used for decades to come. It is an open forum that unites over 700 members from 50 countries, representing administrations, network operators, manufacturers, service providers, and users. Any European organisation demonstrating an interest in promoting European telecommunications standards has the

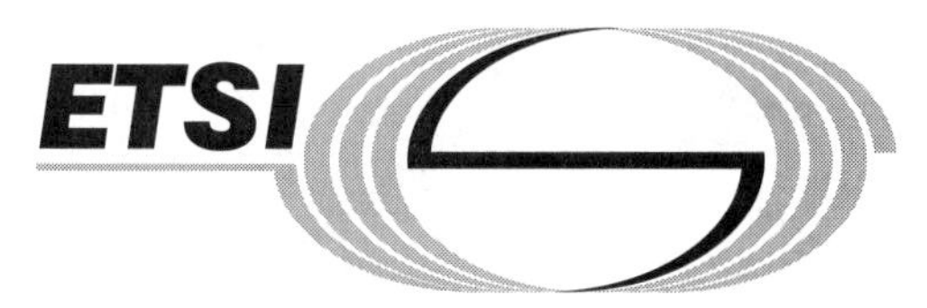

Figure 1.20 ETSI logo

right to represent that interest in ETSI and thus to directly influence the standards making process.

ETSI's approach to standards making is innovative and dynamic. It is ETSI members that fix the standards work programme as a function of market needs. Accordingly, ETSI produces voluntary standards – some of these may go on to be adopted by the EC as the technical base for Directives or Regulations – but the fact that the voluntary standards are requested by those who subsequently implement them means that the standards remain practical rather than abstract.

ETSI promotes the world-wide standardisation process whenever possible. Its Work Programme is based on, and co-ordinated with, the activities of international standardisation bodies such as ISO and IEC.

ETSI consists of a General Assembly, a Board, a Technical Organisation and a Secretariat. The Technical Organisation produces and approves technical standards. It encompasses ETSI Projects (EPs), Technical Committees (TCs) and Special Committees. More than 3500 experts are at present working for ETSI in over 200 groups.

The central Secretariat of ETSI is located in Sophia Antipolis, a high tech research park in southern France. It comprises about 100 staff members. In order to promote and accelerate standardisation, additional experts work on a full-time basis at the ETSI Headquarters. At present, there are about 30 Specialist Task Forces (STFs) with around 100 experts. Up to now, over 2800 ETSI deliverables have been published.

1.11.6 How a Standard is made

1.11.6.1 Drafting

There are several ways to start making a harmonised standard:

- an initial document comes from the International Electrotechnical Commission (80 per cent of cases);
- a document of European origin arises in one of CEN/CENELEC/ETSI's own technical bodies;
- a first draft of a European document comes from one of CEN/CENELEC/ETSI's co-operating partners;
- the National Committees themselves.

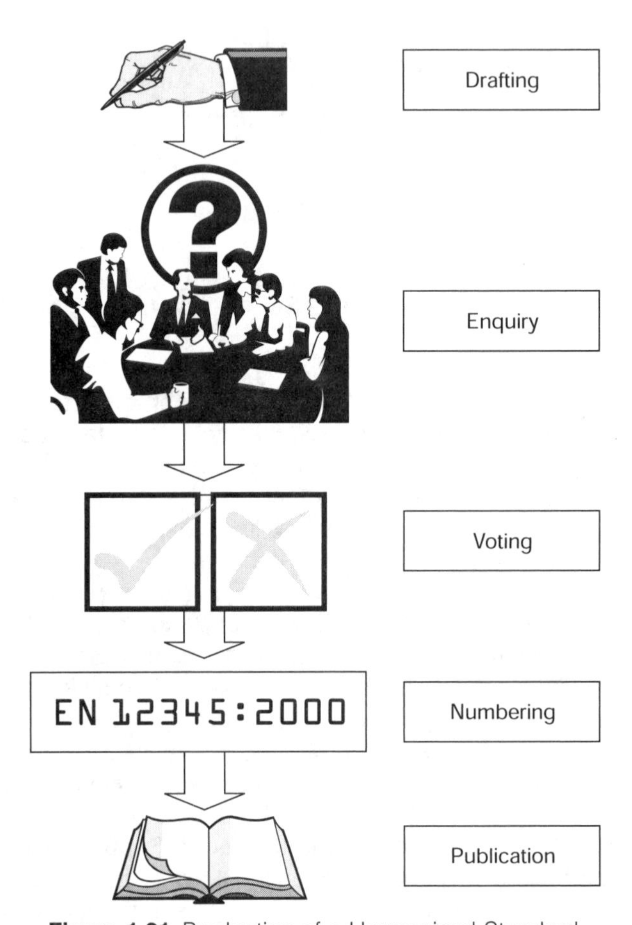

Figure 1.21 Production of a Harmonised Standard

New initiatives originating in Europe are offered to ISO/IEC with the request that they be undertaken at international level. Only if IEC does not want to undertake the work, or if it cannot meet CEN/CENELEC/ETSI's target dates, does the work continue at European level.

Once CEN/CENELEC/ETSI has started work by selecting an international standard, or any other document, to develop into a European Standard, all national work on the same subject is immediately stopped. This suspension of national work is called 'STANDSTILL'.

1.11.6.2 Enquiry

When a suitable draft is available, it is submitted to the National Committees for enquiry – a procedure that lasts six months. Then the comments received are studied by the technical body working on the draft and incorporated into the document, where justified, before a final draft is sent out for vote. The vote usually takes three months.

1.11.6.3 Voting

At this stage, the members have weighted votes corresponding to the size of the country they represent. For instance, the larger countries like Germany, France, Italy and the UK have 10 votes each while the smaller ones have one or two weighted votes. There are two requirements for ratification of the standard. The vote must yield:

- a majority of National Committees in favour of the document;
- at least 71 per cent of the weighted votes cast are positive.

1.11.6.4 Numbering

The shortest unambiguous reference to European Standards is to use its number. The number of a European Standard consists of the capital letters 'EN' followed by a space and a number in arabic numerals, without any space. For example:

- EN 50225:1996 (the year of availability of this EN is indicated by a colon);
- EN 50157–2-1:1996 (the part number is separated by a hyphen).

The first two numerals indicate the origin of the standard:

- 40000 to 44999 cover domains of common CEN/CENELEC activities **within** the IT field;
- 45000 to 49999 cover domains of common CEN/CENELEC activities **outside** the IT field;
- 50000 to 59999 cover CENELEC activities;
- 60000 to 69999 refer to the CENELEC implementation of IEC documents with or without changes.

The IEC and the ISO have allocated themselves blocks of publication numbers: from 1 to 59999 for the ISO and from 60000 to 79999 for the IEC.

1.11.6.5 Publication of Harmonised Standards

Once the harmonised standards have passed the formal acceptance procedures, a list of references has to be published in *OJEC*.

Details of the national standards transposing the harmonised standards then also needs to be published (in the *Official Journal*) by the Member State concerned.

Historically, the majority of European product safety, machinery, and EMC standards are traceable back to their German standards origin and thousands of German standards (DIN, VDI, VDE, VBG, ZH, etc.) form the basis for the newer harmonised standards (EN, IEC, etc.).

Whilst it is generally understood, however, that harmonised standards are the 'minimum safety criteria', a product's compliance with the harmonised standards does not necessarily guarantee a safe product! The manufacturer may have to exceed these requirements in order to make a truly safe product – and one that also meets the customer's expectations. Doing anything less may put the consumer at risk and increase the manufacturer's liability for damages.

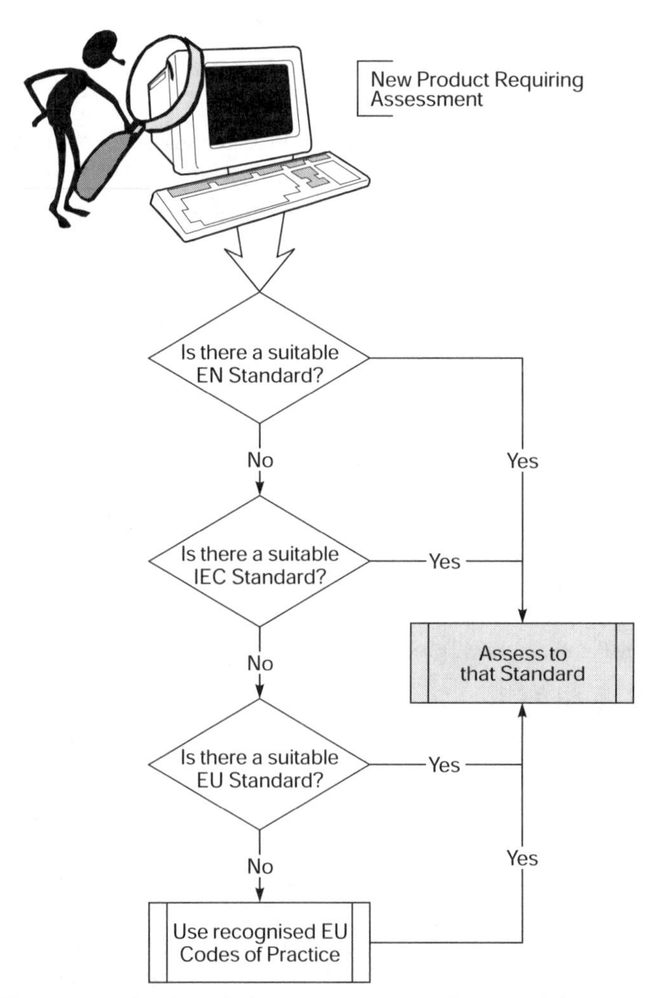

Figure 1.22 Product Safety Assessment – Choice of Standards

When considering standards for a product safety assessment the following list (in order of preference) should be used:

- a European harmonised standard (EN), published in the *Official Journal*;
- an IEC standard that has been published in the relevant Directive(s) (when an 'EN' does not exist);
- an EU national standard (when a harmonised EN or IEC does not exist);
- codes of practice recognised within the EU (where no EN, IEC, ISO or National Standard exists).

1.11.7 European Standards (ENs)

ENs are established as a general rule because it is important that members' national standards become identical wherever possible. Members are obliged to implement European Standards by giving them the status of national standard. ENs are adopted after a formal weighted vote where 71 per cent or more of the expressed votes are in favour. (Countries voting against are still obliged to adopt such standards.)

1.11.8 European Pre-Standards (ENVs)

ENVs are established as prospective standards for provisional application in technical fields where the innovation rate is high, or where there is an urgent need for guidance, and primarily where the safety of persons or goods is not involved.

ENVs do not have to be adopted by the members (but they must be announced and made available).

1.11.9 CEN Reports (CRs)

In order to provide information, the Technical Board of CEN may adopt a report by a simple majority decision. A similar service is also available for CENELEC and ETSI.

1.11.10 CEN Workshop Agreements (CWAs)

CEN Workshop Agreements (CWAs) are consensus-based specifications, drawn up in an open workshop environment. There are similar publications from CENELEC and ETSI, and whilst not being formal standards, they provide interim information for Member States.

1.12 Recognition of European Standards

With the advent of 'New Approach Directives', Europe's product standards and assessment procedures are now recognised by most countries as world-wide standards, making total harmonisation a truly global approach.

Since most ENs are based on IEC/ISO standards, compliance with the EN standards also ensures conformity with equivalent IEC standards. The world is quite obviously following Europe's lead, and most countries have, or eventually will have, accepted the European standards or their IEC equivalents. Thus it is true to say that, meeting European standards will also help manufacturers to comply with technical standards world-wide.

The requirements contained in the European standards and the level of testing required of the European bodies are recognised as far stricter than their US counterparts. It is generally acknowledged that products meeting the European standards exceed US standards, and a product that only meets the US standards almost certainly does not comply with the European Union requirements. This is generally true in most areas of product and component safety, machine safety, and EMC. Concerning safety, there are numerous technical differences between US and EU standards with conflicting guidelines for components construction, fault and ground tests, labelling, manuals, documentation, and warnings. Regarding EMC, the test configuration and limits vary between the United States and the European Union. Most notably, Europe requires additional immunity tests and needs coverage of more product categories. Another difference is that the US focuses primarily on 'flammability and testing', whereas the EU focuses on 'shock hazards and construction'.

Thus, although both the USA and EU safety experts consider the same safety aspects and tests, it is with a different emphasis.

But why do the United States and European Union place a different emphasis on safety hazards? 'Fire' hazards are the most important safety consideration in the United States and 'shock' hazards are of the highest priority in Europe.

The underlying reason for this is to do with the difference in voltage and current (i.e. the mains' power supplied to products). For example, the voltage in the United States is about 115 volts for portable products, with double the input current (i.e. 6 amps), whereas in Europe the voltage is

EU	*USA*
shock	fire
construction	testing
office/industrial environment (i.e. clean/dust and water)	outside/inside environment (rain/no-rain);
high potential and ground testing	high potential or no-tests
'approved' liability	'tested' liability

Table 1.3 Contrast between the EU and US safety views

Figure 1.23 High current versus high voltage!

double that of the United States (i.e. about 230 volts) with half the current (i.e. 3 amps). Hence, with the higher voltage in Europe there is an increased risk of shock. With double the current in the United States the risk is primarily fire.

It cannot be overemphasised that European and international standards (EN and IEC/ISO) are fast becoming the *de facto* world-wide requirements for product safety, EMC, quality and environment management.

1.13 The CE Conformity Marking Directive

The overall aim of the CE Directive is to:

- symbolise that a product complies with the level of protection stated in the harmonisation Directive(s);
- indicate that the manufacturer has undergone all the evaluation procedures laid down by EU law in respect of his product.

This Directive lays down the rules for affixing the CE Marking during manufacture, placing on the market, entry into service and use with industrial products. It also states that:

> The procedures for conformity assessment which are to be used in the Technical Harmonisation Directives relating to the marketing of industrial products will be chosen from among the modules listed in the Annex (i.e. the Annex to the Directive) and in accordance with the criteria set out in this Directive and in the general guidelines in the Annex.

In the main body of the Directive, the Council indicates that:

- limited flexibility is permitted for the inclusion of additional modules (or variations of those modules) to the ones contained in the Directive, when the specific circumstances of a particular Directive so warrant;
- conformity should be assured without imposing unnecessary conditions on manufacturers, and by means of clear and comprehensible procedures;
- manufacturers should ensure that products are in full conformity with the Essential Requirements laid down in the Technical Harmonisation Directives, in particular, the health and safety of users.

The Directive was subject to a four-year transitional period during which the Commission had to:

- report back on any special problems raised by the incorporation of Council Directive 73/23/EEC (i.e. the Low Voltage Directive) of 19 February 1973 – relating to electrical equipment designed for use under certain voltage limits – within the scope of CE Marking procedures, and, in particular, whether safety was being compromised;
- review any problems raised by the issue of overlapping Council Directives, and whether any further Community measures were required.

The Directive finally became effective during 1997 with the overall aim of ensuring that all industrial products that are placed on the market do not compromise the safety and health of users when properly installed, maintained and used in accordance with their intended purpose. Users and third parties should, therefore, be provided with a high level of protection and the devices should attain the performance levels claimed by the manufacturer.

CE Marking in itself is not just about quality or safety, nor is it intended to convey any meaning to consumers who are, after all, reasonably entitled to assume that what they buy from a reputable source is legal, safe and capable of satisfying their requirements. The prime aim of the CE Conformity Marking Directive (indeed, of all New Approach Directives) is to free up the markets within the European Community. But to facilitate free trade, there is a need to identify those industrial products that conform to the requirements of the Directive. In this way an industrial product that has been checked for conformance in one EU country (and has been identified as such) can be sold in any other EU country – **without having to be rechecked**.

With this in mind, it has been agreed (by the Council of Ministers) that all industrial products (other than those that are custom-made or used for clinical investigations), shall bear the CE Marking to indicate their conformity with the provisions of the Directive and to enable them to move freely within the EU.

CE Marking is, therefore, a mandatory conformity marking which shows the compliance of products with the provisions of over 20 Directives which relate to safety, public health, consumer protection, or other Essential Requirements of EU interest.

1.14 CE Marking

CE Marking is the European proof of conformity and is also described as a 'passport' that allows manufacturers and exporters to circulate products freely within the EU. The letters 'CE' (for the French 'Conformité Europeénne') indicate that the manufacturer has satisfied all assessment procedures specified by law for its product.

Figure 1.24 CE Marking

Although consumers may perceive CE Marking as a quality mark, it is not. CE Marking addresses itself primarily to the national surveillance authorities of the Member States, and its use simplifies their task. Just looking at the CE Marking will not tell surveillance authorities to which Directive a given product complies. Rather, it is the declaration of conformity that contains the details of the Directive(s) to which the product complies and the standards that were relied upon in assuring compliance.

The CE Marking must be affixed to the product, to its data plate or, where this is not possible or not warranted on the account of the nature of the product, to its packaging, if any, and to the accompanying documents by the manufacturer, the authorised representative in the EU or, in exceptional cases, by those responsible for placing the product on the market. CE Marking must be affixed visibly, legibly and indelibly. Where special provisions do not impose specific dimensions, the CE Marking must have a height of at least 5 millimetres. CE Marking:

- applies to all equipment put into service in a public or private capacity for professional or non-professional use, or paid for or free of charge;
- is the product manufacturer's symbol of self-declaration;

- indicates a product's conformity to the minimum requirements of the applicable Directives;
- is **not** an approval mark, certification, or quality mark;
- **is** a 'symbol' of the manufacturer's declaration of conformity that implies conformity with the 'minimum requirements' set out in the Directives.

There is no such thing as 'CE approval' or 'CE certification'! CE is not a mark of approval; it is a **marking** which is a manufacturer's self-declaration of conformity.

The main goals of the CE Marking are, therefore, to:

- indicate a product's conformity to the 'Essential Requirements' of the Directives;
- allow products to be 'placed on the market';
- ensure the 'free movement of goods'; and
- permit the 'withdrawal of non-conforming products' by customs and enforcement authorities.

1.15 Directives Affected by the CE Conformity Marking Directive

The CE Conformity Marking Directive (93/465/EEC) is an all-encompassing document covering the conformity assessment of all New Approach Directives.

Although Directives do not necessarily include health and safety or work practice regulations (as they mainly concentrate on regulations that affect industrial product design), clearly there are occasions when an industrial product will need to meet more than one Directive.

For completeness, therefore, Table 1.4 also shows the Directives that could apply to a company manufacturing industrial products.

As agreed by the Council of European Communities, each EC Member State is allowed to convert Directives into national law. For example in the UK, the 'Machine Safety Directive' (MSD) has been adopted as the 'Supply of Machinery (Safety) Regulations 1992'. Obviously there may be some differences between the regulations of one country with another, because of the way in which they have been adopted. In **all** cases, however, the original Directive will always take precedence and the manufacturer and end user must **always** refer to this document in the final instance. **All** Directives also include a requirement to maintain compliance following initial assessment and to achieve this, some form of quality procedures are required.

During a Directive's transitional period (see 1.2), a manufacturer is allowed to choose whether or not to meet the Directive at that time. When this happens, the manufacturer has to state the reasons why he is **not**

Directive	*Ref No.*	*CE in force*
Low-voltage Directive (LVD)*	73/23/EEC	1/1/1997
Simple pressure-vessels*	87/404/EEC	7/1/1992
Safety of Toys*	88/378/EEC	1/1/1990
Construction industrial products*	89/106/EEC	6/27/1991
Electromagnetic compatibility (EMC)*	89/336/EEC	12/31/1995
Personal protective equipment*	89/686/EEC	7/1/1995
Non-automatic weighing machines*	90/384/EEC	1/1/93
Active Implantable Medical devices (AIMD)*	90/385/EEC	12/31/1994
Gas appliances	90/396/EEC	12/31/1995
Hot Water Boilers*	92/42/EEC	1/1/1988
Explosives for civil use	93/15/EEC	1/1/2003
Medical devices – (MDD)	93/42/EEC	6/15/1998
Equipment for use in explosive atmospheres	94/9/EC	6/30/2003
Recreation craft (small boats)	94/25/EC	6/16/1998
Passenger lifts*	95/16/EC	7/1999
Medical devices – in vitro diagnostics (IVD)	COM(95)130	7/1/2002
Household appliances (energy efficiency)	96/57/EC	3/9/1999
Pressure equipment	97/23/EC	5/29/2002
Machinery	98/37/EC	12/1/1998
Radio & Telecommunications Terminal Equipment	99/5/EEC	12/02/98

Notes
* Council Directive 93/68/EEC (dated 22 July 93) gave details of how to amend these Directives for conformance following the change from 'EU Mark' to 'CE Marking'. Directive reference numbers ending with EC were issued after 1993.

Table 1.4 European Directives that are covered by the CE Conformity Marking Directive

currently meeting the Directive. He can, however, still apply the CE Marking so long as it meets the requirements of the other associated Directives and provided that the reasons have been included in his industrial product documentation.

A guide called 'Getting a Good Deal in Europe' outlines how the UK Government is working to cut red tape in Brussels. The Guide may be obtained from the European Community Section of the DTI's (Department of Trade and Industry) Deregulation Unit (Tel: 0171 215 6394).

1.16 Quality Marking

CE Marking is not yet another type of quality marking, although it is often wrongly perceived as such and then frequently compared to other quality marks.

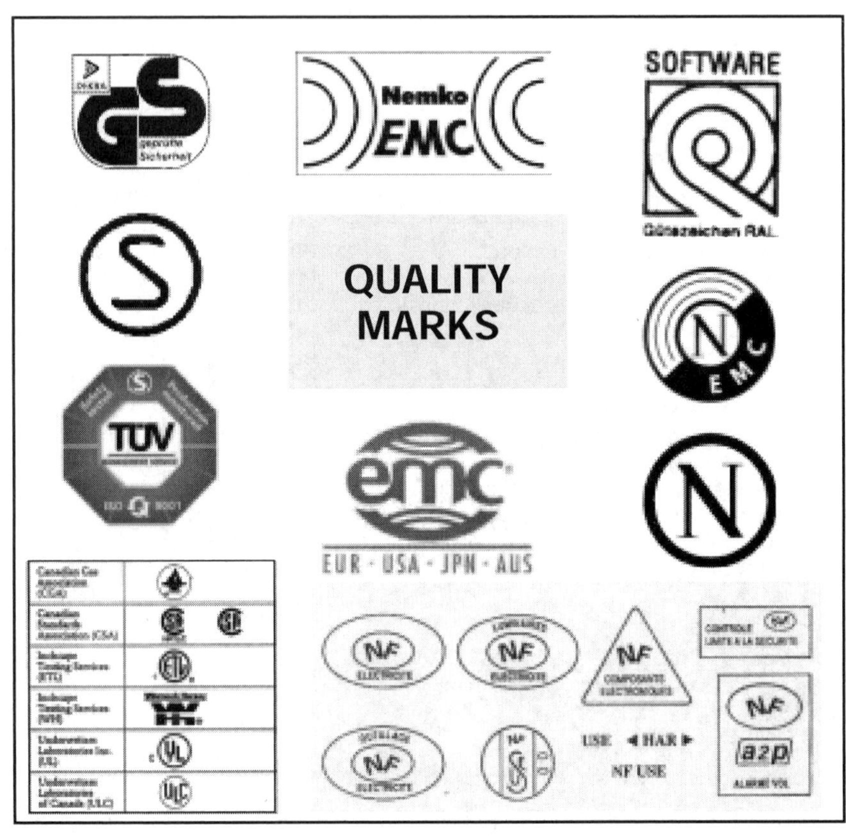

Figure 1.25 Examples of Quality Marks

Certification systems using marks of conformity such as the BSI Kitemark or the BSI Safety Mark, as opposed to the CE Marking, are voluntary, address consumers or users, and tend to influence their appreciation toward the relevant product. Thus, they have a different function to that of the CE Marking.

BSI Safety Mark

Figure 1.26 The BSI Kite and Safety Marks

1.17 Conformity Assessment

The EU has developed several standard modules for conformity assessment and testing of products, either by the manufacturer or through a third party. Each Directive will indicate which module or modules are applicable. Most Directives allow the manufacturer and the exporter to choose a module or combination of modules in order to demonstrate conformity with the Directives. The following modules are available:

- Module A: Internal Production Control;
- Module Aa: Intervention of a Notified Body;
- Module B: EC-Type Examination;
- Module C: Conformity to Type;
- Module D: Production Quality Assurance;
- Module E: Product Quality Assurance;
- Module F: Product Verification;
- Module G: Unit Verification;
- Module H: Full Quality Assurance.

These modules are further explained in Chapter 3.

1.17.1 At Design Level

1.17.1.1 Internal Production Control (Module A)

The procedure is that a manufacturer declares that their product meets the requirements of the Directive(s) concerned and then affixes the CE Marking onto each product, draws up a written Declaration of Conformity and a technical documentation/construction file.

An option (Module Aa) provides a procedure whereby a Notified Body ascertains and attests that a specimen representative of the production planned meets the provisions of the Directive that applies to it. The manufacturer lodges the application with a Notified Body of his choice.

1.17.1.2 EC-Type Examination Certificate (Module B)

Where a specimen representative of the production submitted to an EC-type examination meets the provisions of the Directive, the Notified Body issues an EC-type examination certificate to the applicant. The certificate contains the name and address of the manufacturer, results of the examination, conditions for its validity and the necessary data for identification of the type approved.

1.17.2 At Production Level

1.17.2.1 EC Surveillance (Modules D, E and H)

The purpose of surveillance is to make sure that the manufacturer duly fulfils the obligations arising out of the quality system approved by the Notified Body, and that he maintains it so that it remains appropriate and efficient.

1.17.2.2 EC Verification (Modules F and G)

The purpose of verification is to check and attest through a Notified Body that the products satisfy the provisions of the Directives that apply to them or the model approved by an EC-type examination certificate.

1.17.2.3 Declaration of Conformity (Module C)

Declaration of Conformity is the procedure by which a manufacturer upholds and declares that the product satisfies the provisions of the Directives that apply to them.

Depending on the methods followed for product design acceptance, this procedure can require that production be in conformity with the approved type (i.e. conformity to type) or allow production of any product in conformity with the Essential Requirements.

The declaration of conformity must contain the following information:

- product identification;
- the EU Directives complied with;
- standards used to verify compliance with the Directives;
- name of notified body (not always required, depends on individual Directives);
- signature of or on behalf of the manufacturer or the authorised representative;
- identity of that signatory;
- the manufacturer's name and address.

Most products which are within the scope of European product legislation can be self-certified by the manufacturer and exporter and do not require the intervention of a Notified Body. Certain products may not be self-certified, but must be subjected to the EC-type examination. This examination involves the inspection of a representative example by or on behalf of an external inspection organisation or Notified Body within the EEA.

1.17.3 Notified Bodies

Notified Bodies are independent testing houses or laboratories authorised by the EU Member States to perform the conformity assessment tasks specified in Directives. Manufacturers and exporters may choose a notified body in any EU Member State.

1.17.4 Technical File or Technical Documentation

Most New Approach Directives state that the manufacturer must keep a technical file at the disposal of the relevant national authorities for inspection purposes, or by the Notified Body for EC-type examination.

The Technical Construction File (TCF) demonstrates the technical basis for conformity of the product to the requirements of the Directive. The manufacturer must implement internal measures to ensure that the product remains in conformity. The file is intended essentially for national surveillance authorities.

The TCF must be kept at the disposal of national surveillance authorities for inspection and control purposes, and be available for at least ten years, starting from the production date of the final product. The main elements comprising a TCF are the following:

- Declaration of Conformity containing:
  - a general description of the product;
  - conceptual design;
  - manufacturer's drawings, diagrams of components, sub-assemblies, circuits, etc.;
  - descriptions and explanations necessary for the understanding of the said drawings and diagrams and the general functioning of the product;
  - detailed technical data for essential aspects of the product;
  - a list of the standards referred to and descriptions of the solutions adopted to meet the Essential Requirements of the Directive;
  - results of design calculations made, examinations carried out etc.;
  - test reports;
  - certificate and inspection reports;
  - in the case of series production, the internal conditions that have been observed to safeguard compliance with the Directive.
- CE User Manual.

1.17.5 CE User Manual

The Directives usually have a direct relation to user safety. Information provided to a user plays an essential role in avoiding or reducing safety risks. Thus, a User Manual is often an essential safety requirement. If not available in-house, it is advisable to seek technical and legal assistance in compiling the CE User Manual. A User Manual must contain all the information required for the correct and safe use of a product, including:

- information on risks;
- identification and discouragement of hazardous applications;
- instructions on how the product can be put to safe use;
- details of who is authorised to perform certain actions;
- identification of appropriate safety precautions to be taken.

The CE User Manual must be drawn up in the language of the country/ countries in the EEA into which the product is imported and in the language of the country in the EEA where the product is to be used.

1.18 Products

New Approach Directives apply to products, which are intended to be placed on the European market and/or put into service, for the first time. Normally this means new products but in some circumstances it can also

Product Status	*Mandatory Compliance*	*Possible Compliance*[1]	*Does NOT have to comply*[2]
	New		
		Second-hand	
		Used	
		Modified	
		Combinations of products	
			Repaired

Notes
1. Compliance is dependent upon the relevant product Directive.
2. Subject to the original product complying with the relevant Harmonised Directive and the repair being to the original specification.

Table 1.5 Product Compliance to Harmonised Directives

apply to used and/or second-hand products (e.g. those imported from a third country). Some New Approach Directives explicitly exclude products that are specially designed for military or police purposes and most maintenance operations are also excluded from the scope of the Directives. In all Directives it is left up to the manufacturer to verify whether or not his product is within the scope of a particular Directive.

The actual meaning of a 'product' will vary between the New Approach Directives and can mean equipment, apparatus, devices, appliances, instruments, material, assemblies, components (especially safety components), units, fittings, accessories or systems. In some cases, a combination of different products and parts designed and/or put together by the same manufacturer may be considered as one finished product.

The decision as to whether a combination of products and parts needs to be considered as one finished product, therefore, needs to be taken on a case-by-case basis. Even products which have had their performance, purpose and/or type modified after they have been put into service may be considered as a new product. In these cases it is the person who carries out important changes on the product who is responsible for verifying whether or not it should be considered as a new product.

Products which have been restored (e.g. following a defect) without changing the original performance, purpose and/or type, are not to be considered as new products according to the New Approach Directives.

1.19 Industrial Products

By definition, the placing on the market of an **industrial product** is, as a general rule, governed by the CE Conformity Marking Directive. On the other hand, the placing on the market of a **medicinal industrial product** is governed by Directive 65/65/EEC (relating to the regulations and administrative requirements for proprietary medicinal products). If, however, the device **and** the medicinal industrial product form a single integral unit which is intended exclusively for use in the given combination and which is not reusable, then that single unit industrial product shall be governed by Directive 65/65/EEC.

1.19.1 Industrial Product Withdrawal

Under the terms of the CE Directive, manufacturers of devices which may compromise the health and/or safety of users by:

- failing to meet the Essential Requirements of a Directive;
- incorrect application of the standards referenced by a Directive;
- shortcomings in the standards themselves;

'are required to take all appropriate interim measures to withdraw such devices from the market'. They are also required to prohibit or restrict their being placed on the market and/or put into service. The reasons for having to remove these devices and why non-compliance with the Directive has occurred will have to be stated in the manufacturer's documentation.

1.19.2 European Product Liability

Since the introduction of European product legislation and CE Marking, product liability has become a very important factor for the manufacturer, exporter or importer in conducting trade in Europe. The central element of European product legislation is user safety.

If damage or injury is sustained from a particular product, the user may hold the manufacturer, authorised representative, agent or importer responsible. European legislation in the area of product liability allows users of products to claim damages as the result of an injury. All manufacturers, exporters and importers have a responsibility to ensure that the products they place on the market are safe. European product liability legislation provides instruments and enforcement for users who want to claim damages.

It is therefore necessary for parties involved in the placing of products on the European market to reach contractual agreements in order to cover liability issues.

Liability aspects should be covered not only in the CE Marking administrative requirements such as the declaration of conformity, Technical Construction File and user manual, but also in the sales contract.

1.19.3 CE Marking and ISO 9000

Companies that have in their possession a quality management certificate demonstrate with this that they have an efficient organisational form and that they have low failure costs. Failure costs are the costs incurred due to organisational shortcomings within the company. The quality system makes no reference to the quality of the product. The quality certificate is, however, only a recommendation for customers that their order will be processed correctly and on time. In a similar manner, CE Marking only indicates that the product complies with the Essential Requirements relating to safety, health, environment and consumer protection of the user.

Some Directives explicitly make use of a quality management system (i.e. ISO 9000) as part of the conformity assessment. Only in specific cases is this a requirement to comply with CE Marking Directives.

If a company wishes to provide the customer with assurance about the functional quality of the product, the company needs to obtain a voluntary quality inspection mark, such as Kitemark in the UK and the UL in the USA. The inspection mark guarantees that products will conform to safety and functional requirements over the long term. It is sometimes also the case that these products meet the Essential Requirements of European Directives. In such cases, little extra effort is required to meet CE Marking requirements.

2 STRUCTURE OF NEW APPROACH DIRECTIVES

The European Union's 'New Approach' Directives (often known as 'CE Marking' Directives), have a major impact upon a wide range of industries, particularly the electrotechnical sectors. The underlying principle behind the Directives is the free movement of goods.

The principles and main elements that make up the body of New Approach Directives are contained in a series of Articles (contained in the body of the text) and a number of supporting Annexes.

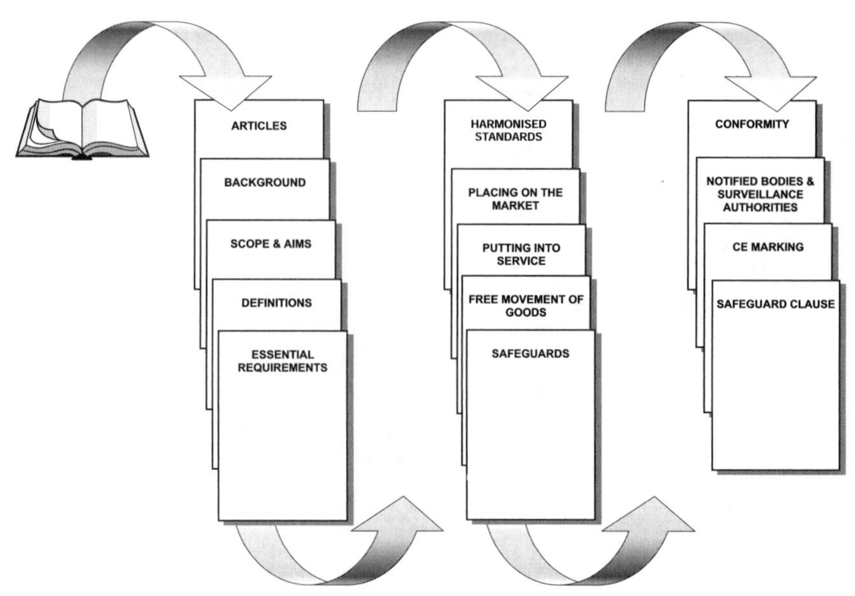

Figure 2.1 Directives, harmonised standards and conformity marking

2.1 Articles

The body of the text of the Directive will contain (depending on the context of the Directive) a number of Chapters each with a number of Articles. These will cover:

2.1.1 Background

This contains an overall statement that 'Products covered by the Directives may be placed on the market only if they do not compromise the safety of persons, domestic animals or property'. Criteria of destination, proper installation, and foreseeable use of a product may be cited in this clause (Note: Member States may impose national regulations (to ensure, for example, that workers are protected provided that the products are not modified in ways unspecified in the Directives)).

The following details are also covered:

- the broad scope of the Directive and its basic requirements;
- previous Directives;
- the requirements of the Treaty establishing the European Community (in particular, Article 100a);
- the harmonised standards and their management;
- the requirements for Member States to exchange information between Notified Bodies;
- the withdrawal of non-compliant products;
- the review and functioning of the Directive.

Having established the overall mandate for adopting a particular New Approach Directive, the document then goes into specific requirements. These are divided into a number of Chapters (depending on the Directive) and a number of associated Annexes. These cover scope and aims, definitions and Essential Requirements, as outlined below.

2.1.2 Scope and Aims

The scope defines the range of products covered and the type of hazards that the Directive is intended to avert. It usually covers risks related to a product (i.e. product approach) or risks related to a phenomena (i.e. risk approach). It may also list the products which are excluded from the Directive. It should be noted that several Directives might cover a particular product.

2.1.3 Definitions

Details of some of the most important terms and definitions used in the Directive are listed in this Article.

2.1.4 Essential Requirements

Essential Requirements lay down the necessary elements for protecting public interest. They are either contained in the body of the Directive or are in one of its annexes and represent the core of European Union law. They are mandatory requirements and are aimed at protecting the health and safety of users (e.g. consumers and workers) and, on some occasions, the protection of property and/or the environment.

The Directive's Essential Requirements are designed to provide and ensure a high level of protection from certain risks (i.e. hazards) associated with a product (such as physical and mechanical resistance, flammability, chemical, electrical or biological properties, hygiene, radioactivity, accuracy). They can either refer to the product or its performance and can contain conditions regarding materials, design, construction, manufacturing process, instructions drawn up by the manufacturer and/or stipulate the principal protection objectives.

As Essential Requirements are applied as a function of the product's risk (i.e. hazard), manufacturers need to carry out a hazard analysis in order to determine the essential requirement that is applicable to that particular product. This analysis should be fully documented in the manufacturer's technical file.

Essential Requirements will include everything necessary for the protection of the public interest (e.g. product safety, protection of workers and consumers, health and environmental protection). They will be worded so that, on transposition into national law, they are capable of forming a legally binding set of obligations which can be enforced, and which will enable Notified Bodies to certify products as being in conformity with those requirements – even in the absence of standards. Compliance with the Essential Requirements authorises the placing of goods on the market.

Whilst the Essential Requirements define the results to be attained, or the risks and hazards that have to be dealt with, they do not specify or predict the technical solutions for doing so. This is achieved by transposing the Essential Requirements into national legislation, producing a legally binding set of obligations (that can be enforced) through the production of harmonised standards. This flexibility allows manufacturers to choose the best way of meeting the requirements.

Products may be placed on the market and put into service **only** if they are in compliance with the Essential Requirements.

2.2 Harmonised Standards

Under the New Approach, national and European standards are judged in relation to their requirements. Any standard which does not fulfil the requirements cannot be used for the purposes of a Directive. The

Directives emphasise this fact and state that, in order to facilitate the proof of conformity with the Essential Requirements, it is necessary to have harmonised European Standards.

2.2.1 Placing on the Market

Placing on the market means when a product has been transferred by sale, loan, hire, lease or gift. This term 'placing on the market' has been defined in only a few of the Directives. According to the Directive on Toys (88/378/EEC) it covers both the sale and its distribution, free of charge. According to the Directives relating to Medical Devices (93/42/EEC) and In-Vitro Diagnostic Medical Devices (COM(95)130) it means 'making available in return for payment (or provided as a free service) a device, with a view to distribution and/or use, regardless of whether it is new or fully refurbished'.

By agreement, however, a product is 'placed on the market' when it was first made available to the EU. This is considered to have taken place when the product is transferred from the manufacturer (or his authorised representative) to a supplier for distribution and/or use within the EU (no matter whether it was manufactured as an individual unit or as a series production). The transfer may also take place directly from the manufacturer to the final consumer or user.

Placing on the market is considered **not** to take place where a product is:

- transferred from the manufacturer located in a third country (i.e. a country outside of the EU);
- transferred to a manufacturer for further processing (e.g. for modification);
- not (yet) granted release for free circulation by customs (e.g. transit, warehousing or temporary importation);
- in a free zone;
- manufactured in a Member State with a view to exporting it to a third country;
- displayed at trade fairs, exhibitions or demonstrations.

Member States are:

- required to ensure that only products that fully comply with the Essential Requirements of a Directive are allowed to be placed on the market;
- required to ensure that products comply with the provisions of the applicable New Approach Directives and other Community legislation when they are put into service;
- required not to prohibit, restrict or impede the placing on the market and putting into service of products that comply with the applicable New Approach Directive(s);

- obliged to take the necessary measures to ensure that products are placed on the market and put into service, only if they do not endanger the safety and health of persons, or other public interests covered by the Directive, when properly installed, maintained and used for the intended purposes.

Member States are allowed, however, (in compliance with Article 30 and 36 of the Treaty) to adopt additional national provisions to protect workers, consumers or the environment, **provided** that these provisions do not require the product to be modified and/or influence the conditions for placing the product on the market.

Putting into service (which is not usually defined in the Directives) takes place at the moment of first use of the product within the EU by the end user.

2.2.2 Putting into Service

Member States accept the free movement of the products where manufacturers comply with the provisions of a Directive and the demands of its Essential Requirements.

2.2.3 Free Movement of Goods

Member States are not allowed to restrict or hinder the placing on the market or putting into service of products bearing CE Marking, unless the provisions relating to CE Marking are incorrectly applied.

By agreement, Member States are required to presume that **all** products bearing CE Marking comply with **all** provisions of the applicable Directives. They can only prohibit, restrict or impede the free movement of products bearing CE Marking if a risk is not covered, or fully covered, by a particular Directive.

2.2.4 Safeguards

If a Member State finds out that a product with CE Marking does not offer the safety level required, or that the product is not in conformity with the requirements of the Directive, they are required to immediately inform the Commission. The Commission will then review the harmonised standard (for any shortcomings) and if necessary, prohibit or restrict its placing on the market and/or order its removal from the market of the product.

2.3 Conformity

The conformity of products to the Essential Requirements of the Directive can generally be confirmed and attested by a third party chosen by the manufacturer from amongst their country's Notified Bodies.

Products that comply with national standards transposing harmonised standards (the reference numbers of which have all been published in the *Official Journal of the European Communities*), are presumed to comply with the corresponding Essential Requirements. Where the manufacturer has not applied, or has only partially applied such a standard, he must document the measures taken and their adequacy in order to comply with the Directive's Essential Requirements.

The procedure(s) for conformity assessment Directives are chosen from modules listed in the Directive's Annexes.

2.3.1 Notified Bodies and Surveillance Authorities

Third-party conformity assessment is normally carried out by Notified Bodies. These bodies are designated by the Member States and can prove their conformity with EN 45000 series covering the general criteria for the operation and testing of various types of bodies (see Chapter 5), by submitting an accreditation certificate or other documentary evidence.

The Commission publishes and updates lists of Notified Bodies in the *OJEC*.

2.3.2 CE Marking

The requirements for affixing CE Marking to products that comply with the Essential Requirements of a particular Directive will cover how and where the CE Marking is to be affixed plus any additional information that has to be included. It will also indicate the action that Member States should take against wrongly affixed CE Markings.

2.3.3 Safeguard Clause

Member States are obliged to take all appropriate measures to prohibit or restrict the placing on the market of products bearing CE Marking (or to withdraw them from the market) if these products could compromise the safety and health of individuals.

2.4 Adoption of New Approach Directives

New Approach Directives are adopted according to the co-decision procedure provided for in Article 189b of the Treaty. New Approach Directives that have been adopted are published in the L series of the *OJEC* whilst Commission proposals for New Approach Directives are published in the C series of the *OJEC*.

The Commission initiates the adoption of New Approach Directives, or their modification, by making a proposal to the Council and to the

European Parliament. Proposals that have anything to do with health, safety, environmental protection and/or consumer protection must, according to Article 100a, take as a base a high level of protection.

When the Commission receives a proposal, the Council first asks the Parliament and the Economic and Social Committee for an opinion before stating its common position on the proposal. Once an agreed common position has been reached, it is given to Parliament, who must accept, reject or propose amendments during this second reading.

The Commission re-examines its proposal in the light of Parliament's amendments, and returns the proposal to the Council, which has to take a final decision within three months. If necessary, problems are solved in a conciliation committee between the Council and the Parliament, in which the Commission participates as a moderator. The flowchart in Figure 2.2 shows the adoption procedure according to Article 189b of the Treaty.

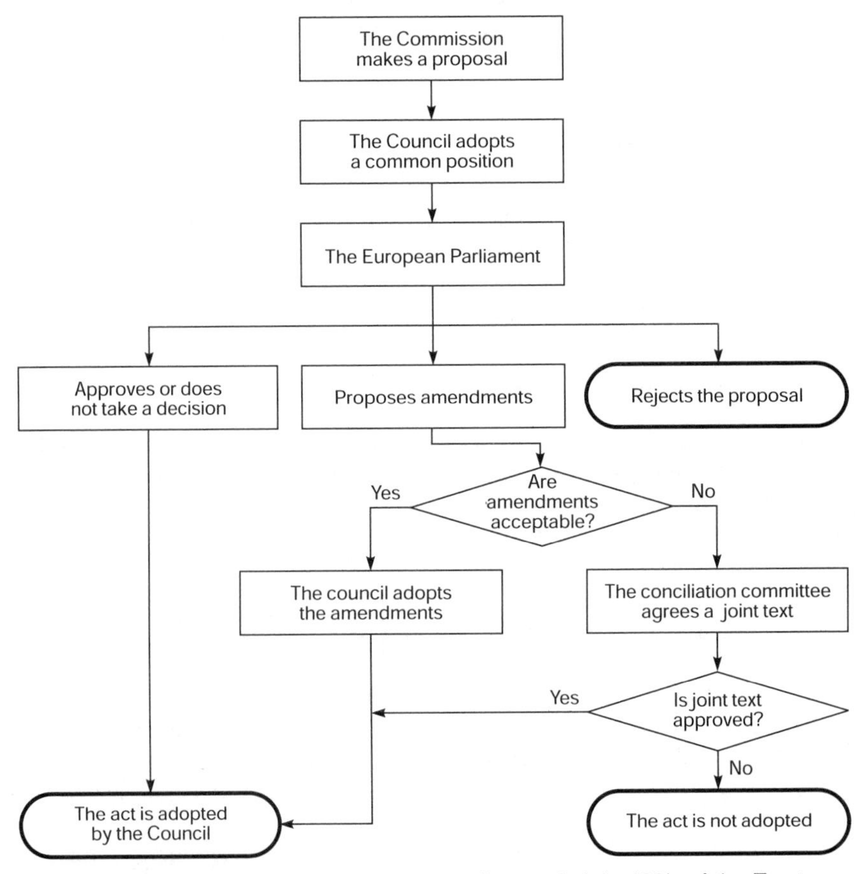

Figure 2.2 Adoption of Directives according to Article 189b of the Treaty

2.5 Transposition of New Approach Directives

New Approach Directives are total harmonisation Directives, which means that they replace all corresponding national provisions. Member States are obliged to transpose them into their own national laws, regulations or administrative provisions. Failure to complete this transposition within the period laid down for that purpose constitutes a serious breach of EU law.

2.6 Standing Committee

The Commission is assisted by an advisory committee (called a 'Standing Committee') who oversees the implementation of a particular Directive.

Any question regarding the implementation of a Directive may be submitted to this Committee, and they must be consulted (in particular) prior to the publication of any transitional standards that Member States may be considering. The rules for the Standing Committee often correspond to the rules set up by the Directive 83/189/EEC relating to the 'Provision of information in the field of technical standards and regulations'.

2.7 Review and Reporting

The Commission is required to regularly review the operation of a Directive. Usually this is completed on a five-yearly basis but it can vary according to individual Directives.

2.8 Transitional Provisions

The aim of the transitional period is to allow manufacturers and Notified Bodies to gradually adjust to the conformity assessment procedures that have been set up by the new Directives. During this period, Member States are obliged to maintain their national systems, unless otherwise detailed in the relevant Directive.

During the transitional period of a Directive the manufacturer usually has the choice to either meet the requirements of the directive or the relevant national regulations, if different. The option chosen, and hence the extent of the conformity expression enshrined in the CE Marking, must be indicated by the manufacturer in the EC declaration of conformity, and in the notices or instructions accompanying the product or, where appropriate, on the data plate.

Member States are not allowed to make any changes (except in cases of *force majeure*) to their national system in question, that might result in a modification of the product requirements, the conformity assessment procedure, or which would otherwise have an effect on acquired rights.

At the end of the transitional period Member States are obliged to terminate (i.e. repeal) their national systems, and products manufactured before or during this period in line with the system to be repealed may no longer be placed on the EU market.

Apart from the Essential Requirements of a Directive, Member States may, however, request that products that met the requirements of national regulations that were in place when the Directive was adopted, may continue to be placed on the market.

2.9 Repeal

Details of the Directives that are to be repealed because of the current Directive have to be listed together with a short reason why they are being repealed.

2.10 Entry into Force

This is the date that the Directive is to published in the *Official Journal of the European Communities* and signifies the date that the Directive comes into force.

2.11 Annexes

The annexes to Directives vary according to the product being covered, but in the main they will provide:

- full details of equipment not covered by the Directive;
- the Conformity Assessment Procedures (i.e. details of the modules to be used to assess conformance with the Essential Requirements);
- the minimum criteria to be taken into account by Member Sates when designating Notified Bodies;
- specific CE Marking requirements.

3 STRUCTURE OF THE CE CONFORMITY MARKING DIRECTIVE (93/465/EEC)

Council Decision number 93/465/EEC states that an industrial product placed on the market shall have CE Marking on it. The CE Conformity Marking Directive is 17 pages long and as well as having the normal general guidelines it also provides details of the size of the actual logo, how to affix the CE Marking etc. and then describes the eight modules for conformity and assessment (see Figure 3.1).

The document is subdivided as shown in Table 3.1.

Main part	The aims and the requirements of the Council Directive itself
Annex – Part I	General Guidelines
Annex – Part II	Modules for Conformity Assessment

Table 3.1 Contents List of the CE

3.1 Main Part

On 22 July 1993 the Council agreed the aims for CE Conformity Marking and published their decisions in 93/465/EEC, which stated that the aim of the CE Marking is to:

> . . . symbolise the conformity of a product with the levels of protection of collective interests imposed by the total harmonisation Directives and to indicate that the economic operator has undergone all the evaluation procedures laid down by Community law in respect of his product.

In accordance with these decisions, it was agreed that '. . . all industrial products which are covered by the technical harmonisation Directives of the EU can **only** be placed on the Market after the manufacturer has affixed the CE Marking to them'. (Note: at that time the term used was EC Mark) and that:

> The procedures for conformity assessment which are to be used in the technical harmonisation Directives relating to the marketing of industrial products will be chosen from among the modules listed in the Annex (to the Directive) and in accordance with the criteria set out in this Directive and in the general guidelines in the Annex.

93/465/EEC thus:

- lays down the rules for affixing the CE Conformity Marking and for its design, manufacture, placing on the market, entry into service and/or use of industrial products;
- emphasises that conformity 'should be assured without imposing unnecessarily onerous conditions on manufacturers, and by means of clear and comprehensible procedures';
- introduces limited flexibility regarding the use of additional modules and/or variations in the modules (when the specific circumstances of a particular sector or Directive so warrant);
- ensures that products (in particular those that can affect the health and safety of users/consumers) conform to the Essential Requirements of the various technical harmonisation Directives.

In publishing their decision the Council stated that the Commission shall:

- report back on any special problems raised by the incorporation of Council Directive 73/23/EEC (i.e. the Low Voltage Directive) within the scope of CE Marking procedures – and, in particular, whether safety is being compromised;
- review any problems raised by the issue of overlapping Council Directives – and whether any further Community measures are required.

3.2 General Guidelines

The essential objective of 93/465/EEC is to provide the public authorities with an assurance that products placed on the Market conform to the Essential Requirements contained in EU harmonised Directives, particularly those concerning the health and safety of users and consumers.

The guidelines contained in Part 1 (of the Annex to the Directive) are divided into two sub-sections – those dealing with the use of conformity assessment procedures and those concerning the rules for actually affixing the CE Marking.

3.2.1 Guidelines for the Use of Conformity Assessment Procedures

Harmonised Directives are required to set a range of possible choices so as to provide a high level of safety for a given product or product sector, in particular:

- the relation (and appropriateness) of the modules to the type of products;
- the nature of the risks involved;
- the economic infrastructures of the given sector (e.g. existence/non-existence of third parties);
- the types and importance of production, etc.

3.2.2 Conformity Assessment Modules

Part II of the Directive's Annex shows how conformity assessment can be subdivided into eight modules relating to the design and production phases and requires that a product should be subject to both of these phases before being placed on the market – and then only if the results are positive.

These conformity modules (see Table 3.2) have been designed so as to provide the manufacturer with the widest range possible for ensuring compliance of his product(s) with the requirements of the various applicable Directives.

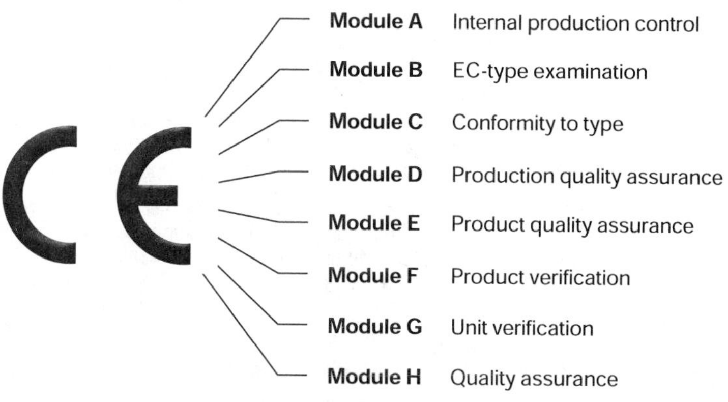

Figure 3.1 CE Marking modules

Module	*Title*	*Short description*
A	Internal Production Control	**Manufacturers declare** that their product satisfies the requirements of the Directive concerned.
B	EC-Type Examination	**Notified Bodies ascertain and attest** that a specimen (i.e. a representative of the production envisaged) meets the provisions of the Directive that applies to it.
C	Conformity to Type	**Manufacturers ensure and declare** that their products are in conformity with the EC-Type Examination certificate(s) that has (have) been issued for it (them).
D	Production Quality Assurance	**Manufacturers declare** that their products satisfy the requirements of the applicable Directive, are in **conformity** with the EC-Type Examination certificate(s) that has (have) been issued for it (them) and that they operate an **ISO 9002:1994** approved Quality Management System.
E	Product Quality Assurance	**Manufacturers declare** that the products concerned satisfy the requirements of the applicable Directive, are in **conformity** with the EC-Type Examination certificate(s) that has (have) been issued for it (them) and that they operate **an ISO 9003:1994** approved Quality Management System.
F	Product Verification	This consists of two parts: **The manufacturer checks and tests to confirm** that the product(s) meet the requirements of the EC-Type Examination certificate(s) that has (have) been issued for it (them) in addition to the requirements of the applicable Directive. **The Notified Body examines and tests** the manufacturer's products for conformity. This can be completed either by examination and/or testing every product or testing products on a statistical basis.
G	Unit Verification	This consists of two parts: **The manufacturer ensures and declares** that the product(s) meet the requirements of the EC-Type Examination certificate(s) that has (have) been issued for it (them) in addition to the requirements of the applicable Directive. **The Notified Body examines and tests** the manufacturer's products for conformity with the relevant requirements of the appropriate Directive.
H	Full Quality Assurance	**Manufacturers declare** that the products concerned satisfy the requirements of the applicable Directive, are in **conformity** with the EC-Type Examination certificate(s) that has (have) been issued for it (them) and that they operate a **full ISO 9001:1994** approved Quality Management System.

Table 3.2 Modules for Conformity Assessment

The applicable Directive will set out the criteria governing the conditions for manufacture and the manufacturer will then choose the most appropriate module(s) for his production. Normally you will find that the Directives avoid imposing unnecessary modules which would be too onerous – however, be careful as this doesn't always happen!

The CE Conformity Marking Directive emphasises that whilst the manufacturer should, whenever possible, be given the possibility of using modules that are based on quality assurance, they should also be given the opportunity of using a combination of modules not using quality assurance, and *vice versa*.

To protect the manufacturers, the technical documentation provided to Notified Bodies is limited to those that are required solely for the purpose of assessment of conformity and the Notified Bodies have been encouraged to apply the modules without unnecessary burden on the manufacturer.

Notified Bodies (see 3.8) are nominated by Member States as being the most competent and technically qualified to provide facilities for the conformity assessment of products to the appropriate Directive. Subcontracting of this work is permitted, but is subject to:

- the competence and capability of the subcontractor;
- their conformity with the EN 45000 series of standards;
- effective monitoring of the subcontractor;
- the ability of the subcontractor to exercise effective responsibility for the work carried out under subcontract.

Notified Bodies who can prove their conformity with the EN 45000 series (by submitting an accreditation certificate or other documentary evidence) are presumed to conform to the requirements of the Directives. The Commission publishes a list of Notified Bodies in the *Official Journal* and this is constantly updated.

3.2.3 Variants of the Basic Modules

Variants of the basic modules are shown in Table 3.3

3.3 Conformity Assessment Modules – short description

Conformity assessment can be subdivided into modules (see Table 3.2) relating to the design and production phases. A product should be subject to both of these phases (i.e. design and production), and requires CE Marking before being placed on the market and then only if the results are positive. There are eight separate modules (see Figure 3.1).

Variant	*Description*	*Additional elements compared to basic modules*
Aa1 and Cbis1	Internal production control, and one or more tests on one or more specific aspect of the finished product	Intervention of a notified body regarding testing, either at design or production stage. The products concerned and the applicable tests are specified in the Directive.
Aa2 and Cbis2	Internal production control, and product checks at random intervals	Intervention of a notified body regarding product checks at production stage. The relevant aspects of the checks are specified in the directive.
Dbis	Production quality assurance without use of module B	Technical documentation is required.
Ebis	Product quality assurance without use of module B	Technical documentation is required.
Fbis	Product verification without use of module B	Technical documentation is required.
Hbis	Full quality assurance with design control	A notified body analyses the design of a product or a product and its variants, and issues an EC design examination certificate.

Table 3.3 Variants of basic modules

3.3.1 Explanatory notes and other considerations

Most of the New Approach Directives incorporate a 'Self-Certification' option. This leaves the responsibility for applying the test standards, and certifying that the products comply, with the product manufacturer. Whilst this means that there is, in these instances, no mandatory requirement for third-party intervention, in practice, many manufacturers will contract part of the assessment process to a third-party laboratory.

For most of the New Approach Directives, the principles of complying are broadly similar:

- products should comply with the applicable Essential Requirements (this is usually a matter of product design and product instructions);
- the manufacturer or supplier should then demonstrate that the product complies with the Essential Requirements using one of the product

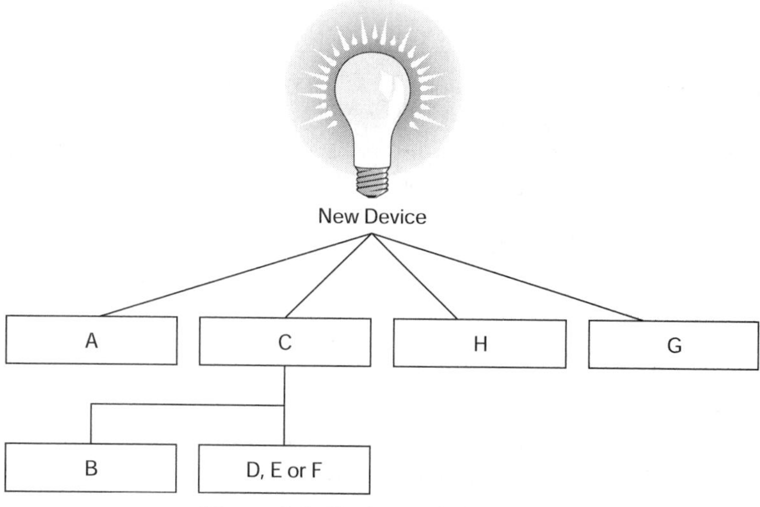

Figure 3.2 Choice of CE modules

conformity assessment modules available within the Directive. (This is usually achieved by assessing the product against the relevant standards and/or preparing the appropriate documentation such as a Technical File);
- CE Marking should be affixed legibly and indelibly to the product;
- the manufacturer should prepare and sign a Declaration of Conformity or the Certificate of Conformity (depending on the Directive concerned).

Specific Directives may use modules A, C and H (i.e. the 'Quality' modules) supplemented by one (or all) of the additional provisions shown (in boxes) in the modules.

Module C is designed to be used in combination with module B (i.e. EC-type examination). Modules D, E and F will also normally be used in combination with module B; however, in special cases (for example, when dealing with certain products of very simple design and construction) they may be used on their own.

3.3.2 Module A

Module A Internal Production Control	Covers internal design and production control. This module does not require a Notified Body to take action.	**Manufacturers declare** that their product satisfies the requirements of the Directive concerned and then affixes the CE Marking onto each product, draws up a written Declaration of Conformity and a technical documentation file.

The manufacturer must establish a technical documentation file and this shall be kept (at the disposal of the relevant national authorities for inspection purposes) for a period ending at least 10 years (specific Directives may, however, alter this period) after the last product has been manufactured.

Where neither the manufacturer nor his authorised representative is established within the EU, the obligation to keep the technical documentation available is the responsibility of the person who places the product on the EU market.

The technical documentation file must enable the conformity of the product with the requirements of the Directive to be assessed and needs to cover the design, manufacture and operation of the product.

As a minimum, the technical documentation must contain:

- a general description of the product;
- conceptual design and manufacturing drawings and schemes of components, sub-assemblies, circuits, etc.;
- descriptions and explanations necessary for the understanding of said drawings and schemes as well as the operation of the product;
- a list of the standards referred to;
- the results of design calculations made, examinations carried out, etc.;
- test reports.

The manufacturer (or his authorised representative) must retain a copy of the Declaration of Conformity with the technical documentation. In addition, he must ensure that the manufacturing process complies with this technical documentation and with the requirements of the Directive that apply to that particular manufacturing process.

Note

The EMC Directive sometimes mandates the use of a Competent Body for mandatory certification and a special file called the technical construction file (TCF). This is the case when no harmonised standards exist or they are applied only in part. The TCF is generated by the manufacturer in conjunction with the Competent Body. This process is commonly called the 'TCF route' and is a certification process that requires a 'Certificate of Conformity' issued by a Competent Body prior to CE Marking. The TCF process is frequently used when there are numerous product variations or for large machines.

3.3.2.1 Module Aa

This module consists of module A plus one of the following options:

3.3.2.1.1 Option 1

For each product manufactured, the manufacturer must carry out one or more tests on one or more specific aspects of the product.

When authorised, the manufacturer must affix the Notified Body's identification number during the manufacturing process.

3.3.2.1.2 Option 2

A Notified Body (chosen by the manufacturer) must carry out product checks at random intervals. An adequate sample of the final products, taken on site by the Notified Body, must be examined and tested in accordance with the relevant standard, and conformity of the product must be checked with the relevant requirements of the Directive. The product check must include:

- the sampling plan;
- operational characteristics, etc.;
- statistical method(s) to be applied.

The manufacturer must affix the Notified Body's identification number during the manufacturing process.

3.3.3 Module B

Module B EC-Type Examination	Covers the design phase, and must be followed up by a module providing for assessment in the production phase. The EC-Type Examination certificate is issued by a Notified Body.	**Notified Bodies ascertain and attest** that a specimen, (that represents the production envisaged) meets the provisions of the Directive that applies to it.

The manufacturer must lodge an application for EC-Type Examination with a single Notified Body. The application must include:

- the name and address of the manufacturer;
- a written declaration that the same application has not been lodged with another Notified Body;
- the relevant technical documentation;
- a representative example of the production envisaged.

The technical documentation must enable an assessment to be made of the conformity of the product with the requirements of the relevant Directive(s) which apply to it. It must cover the design, manufacture and operation of the product and contain:

- a general description of the type of production,
- conceptual design and manufacturing drawings and diagrams of components, sub-assemblies, circuits, etc.,
- descriptions and explanations necessary for an understanding of the said drawings and diagrams and the operation of the pressure equipment;
- a list of the standards, applied in full or in part, and descriptions of the solutions adopted to meet the Essential Requirements of the Directive(s);
- the necessary supporting evidence for the adequacy of the design solution;
- results of design calculations made, examinations carried out, etc.;
- test reports and information about the manufacturing test that have been applied;
- a statement indicating whether or not the product incorporates, as an integral part, a substance which could be considered as an industrial product, in its own right.

The product suppliers must be able to prove, through their documentation, that adequate design, production and quality control procedures were in place to ensure a compliant product, that is without defects.

When neither the manufacturer nor his authorised representative is established in the EU, the obligation to keep available the technical documentation shall fall to the person responsible for placing the product on the EU market, or the importer.

The Notified Body must:

- examine and assess the technical documentation and identify the components which have been designed in accordance with the relevant provisions of the standards referred to;
- assess the materials where these are not in conformity with the relevant harmonised standards;

- perform the necessary examinations to establish whether the solutions adopted by the manufacturer meet the Essential Requirements of the Directive where the standards referred to have not been applied;
- perform the necessary examinations to establish whether, where the manufacturer has chosen to apply the relevant standards, these have actually been applied;
- approve the procedures envisaged for the manufacture of the product;
- verify that the personnel involved in the manufacture of the product are suitably qualified;
- issue an EC-Type Examination certificate to the applicant (where the design meets the provisions of the Directive) containing:
  - the name and the address of the applicant;
  - the conclusions of the examination;
  - conditions for its validity;
  - the necessary data for identification of the approved design.

A list of the relevant parts of the technical documentation must be annexed to the certificate and a copy kept by the Notified Body.

If the Notified Body refuses to issue an EC-Type Examination certificate to the manufacturer, that Body must provide detailed reasons for their refusal and allow for an appeals procedure to be made available.

The applicant must inform the Notified Body that holds the technical documentation concerning the EC-Type Examination certificate of any modifications that have been made to the approved design. These shall be subject to additional approval if these changes affect the conformity of the product with the Essential Requirements of the Directive. This additional approval must be given in the form of an addition to the original EC-Type Examination certificate. Each Notified Body must provide the Member States with details of all:

- EC-Type Examination certificates and additions granted;
- EC-Type Examination certificates and additions withdrawn.

The manufacturer must keep copies of all EC-Type Examination certificates, and their additions, with the technical documentation, for a period of ten years (can vary with Directives) after the last of the product has been manufactured.

3.3.3.1 Module B1

This module is an alternative to the EC-Type Examination and is used when the design of an envisaged product requires approval. The procedure (which is called 'EC-Design Examination') describes how a Notified Body ascertains and attests that the design of an industrial product meets the provisions of the Directive(s) which apply to it.

3.3.4 Module C

Module C Conformity to Type	Covers the production phase and follows module B. Provides for conformity with the type as described in the EC-Type Examination certificate issued according to module B. This module does not require a notified body to take action.	**Manufacturers ensure and declare** that their products are in conformity with the EC-Type Examination certificate(s) that has (have) been issued for it (them) and that they satisfy the requirements of the Directive that applies to it (them).

To achieve this, the manufacturer shall:

- institute and maintain a post-production review system aimed at identifying any malfunction, deterioration of the product's characteristics or performance so that corrective action can be completed. He shall notify the Competent Authority of the following incidents immediately on occurrence:
  - any malfunction or deterioration in any product which may lead to (or have caused) the death of a user (e.g. in the case of the Medical Devices Directive, a patient);
  - any reasons which lead to the systematic recall of products of the same type.
  - affix the CE Marking to each product and draw up a written Declaration of Conformity;
  - keep a copy of the Declaration of Conformity for a period of 10 years (depends on the Directive) after the product has ceased to be manufactured.

Note
When neither the manufacturer or his authorised representative is established in the EU, the obligation to keep available the technical documentation shall fall to the person responsible for placing the product on the EU market, or the importer.

Whilst Module C does not in itself ask for a Quality Management System (QMS), the format of the conformity procedure is such that its documentary requirements are best presented in the form of a QMS.

3.3.5 Module D

Module D Production Quality Assurance	Covers the production phase and follows module B. Derives from quality assurance standard EN ISO 9002, with the intervention of a notified body responsible for approving and controlling the quality system set up by the manufacturer.	**Manufacturers declare** that their products satisfy the requirements of the applicable Directive, are in **conformity** with the EC-Type Examination certificate(s) that has (have) been issued for it (them) and that they operate an **ISO 9002:1994** approved Quality Management System. Manufacturers then affix CE Marking (accompanied by an identification symbol of the Notified Body responsible for the EC monitoring) to each product and draw up a written Declaration of Conformity.

To ensure compliance with this module the manufacturer must operate an approved quality system for production, final product inspection and testing in accordance with ISO 9002:1994. This quality system must ensure compliance of the products with the requirements of the Directive(s) that apply to them.

3.3.5.1 Quality System

The manufacturer must lodge an application for assessment of his quality system with a Notified Body of his choice, for the products concerned. This application must include:

- all relevant information for the product category envisaged;
- the documentation concerning the quality system;
- the technical documentation of the approved type (if applicable) together with a copy of the EC-Type Examination certificate.

All the elements, requirements and provisions that have been adopted by the manufacturer must be documented in the form of written policies, procedures and instructions. This documentation will form the basis of the manufacturer's Quality Management System (QMS) which will include quality programmes, plan, manuals and records.

In particular, a manufacturer's Quality Management System must contain a description of the:

- quality objectives and the organisational structure, responsibilities and powers of the management with regard to product quality;
- manufacturing, quality control and quality assurance techniques, processes and systematic actions that will be used;
- examinations and tests that will be carried out before, during and after manufacture – and the frequency with which they will be carried out;
- quality records, such as inspection reports and test data, calibration data, qualification reports of the personnel concerned, etc.;
- monitoring method to ensure achievement of the required product quality and the effective operation of the quality system.

3.3.5.2 Notified Bodies

The Notified Body will assess the manufacturer's Quality Management System for conformance with ISO 9002:1994. At least one member of the auditing team must have experience of evaluating the product technology concerned and the evaluation procedure must include an inspection visit to the manufacturer's premises.

The Notified Body will also carry out periodic audits to make sure that the manufacturer maintains, applies and duly fulfils the obligations arising out of the approved quality system. The manufacturer must allow the Notified Body entrance (for inspection purposes) to the locations of manufacture, inspection and testing, and storage and must provide the Notified Body with all necessary information, in particular:

- the Quality Management System documentation;
- quality records – such as inspection reports, test and calibration data, qualification reports of the personnel etc.

Under the terms of this module, the Notified Body may also pay unexpected and unannounced visits to the manufacturer to verify and validate that their quality system is functioning correctly.

The manufacturer must keep at the disposal of the national authorities, for a period of ten years (can vary with Directives) after the last of the product has been manufactured:

- the documentation associated with his QMS;
- details of any amendments, alterations or modifications to the QMS;
- previous reports and decisions made by the Notified Body.

Notified Bodies are required to provide other Notified Bodies with the relevant information concerning the quality system approvals that they have issued and withdrawn.

3.3.6 Module E

Module E Product Quality Assurance	Covers the production phase and follows module B. Derives from quality assurance standard EN ISO 9003, with the intervention of a notified body responsible for approving and controlling the quality system set up by the manufacturer.	**Manufacturers declare** that the products concerned satisfy the requirements of the applicable Directive, are in **conformity** with the EC-Type Examination certificate(s) that has (have) been issued for it (them) and that they operate an **ISO 9003:1994** approved Quality Management System. Manufacturers then affix CE Marking (accompanied by an identification symbol of the Notified Body responsible for the EC monitoring) to each product and draw up a written Declaration of Conformity.

To ensure compliance with this module the manufacturer must operate an approved quality system for production, final product inspection and testing in accordance with ISO 9002:1994. This quality system must ensure compliance of the products with the requirements of the Directive(s) that apply to them.

3.3.6.1 Quality System

The manufacturer must lodge an application for assessment of his quality system with a Notified Body of his choice, for the products concerned. This application must include:

- all relevant information for the product category envisaged;
- the documentation concerning the quality system;
- the technical documentation of the approved type together with a copy of the EC-Type Examination certificate.

All the elements, requirements and provisions that have been adopted by the manufacturer must be documented in the form of written policies, procedures and instructions. This documentation will form the basis of the manufacturer's Quality Management System (QMS) and will include quality programmes, plan, manuals and records.

In particular, a manufacturer's Quality Management System must contain a description of the:

- quality objectives and the organisational structure, responsibilities and powers of the management with regard to product quality;
- examinations and tests that will be carried out after manufacture;
- quality records, such as inspection reports and test data, calibration data, qualification reports of the personnel concerned, etc.;
- means to monitor the effective operation of the quality system.

3.3.6.2 Notified Bodies

The Notified Body will assess the manufacturer's Quality Management System for conformance with ISO 9003:1994. At least one member of the auditing team must have experience of evaluating the product technology concerned and the evaluation procedure must include an inspection visit to the manufacturer's premises.

The Notified Body will also carry out periodic audits to make sure that the manufacturer maintains, applies and duly fulfils the obligations arising out of the approved quality system. The manufacturer must allow the Notified Body entrance (for inspection purposes) to the locations of manufacture, inspection and testing, and storage and must provide the Notified Body with all necessary information, in particular:

- the Quality Management System documentation;
- the technical documentation;
- quality records – such as inspection reports, test and calibration data, qualification reports of the personnel etc.

Under the terms of this module, the Notified Body may also pay unexpected and unannounced visits to the manufacturer to verify and validate that the quality system is functioning correctly.

The manufacturer must keep at the disposal of the national authorities, for a period of 10 years (can vary with Directives) after the last of the product has been manufactured:

- the documentation associated with his QMS;
- details of any amendments, alterations or modifications to the QMS;
- previous reports and decisions made by the Notified Body.

Notified Bodies are required to provide other Notified Bodies with the relevant information concerning the quality system approvals that they have issued and withdrawn.

3.3.7 Module F

Module F Product Verification	Covers the production phase and follows module B. A Notified Body controls conformity to the type as described in the EC-Type examination certificate issued according to module B, and issues a certificate of conformity.	This consists of two parts: The **manufacturer checks and tests to confirm** that the product(s) meet the requirements of the EC-Type Examination certificate(s) that has (have) been issued for it (them) in addition to the requirements of the applicable Directive. The **Notified Body examines and tests** the manufacturer's products for conformity. This can be completed either by examination and/or testing every product or testing products on a statistical basis.

The product suppliers must be able to prove, through their documentation, that adequate design, production, and quality control procedures were in place to ensure a compliant product, that is, without defects.

3.3.7.1 The Manufacturer

The manufacturer is responsible for:

- seeing that his manufacturing process ensures compliance of the products with the EC-Type Examination certificate (that has been issued for that product) and the requirements of the applicable Directive;
- affixing CE Marking to each product;
- drawing up a Declaration of Conformity;
- retaining a copy of the Declaration of Conformity for 10 years (depends on Directive) after production has ceased.

Before commencing the manufacturing process, the manufacturer must prepare documents (i.e. Work Instructions) defining the actual manufacturing process (e.g. in the Medical Devices Directive this could be about routines to ensure 'homogenous production and conformity of the products particularly with regard to sterilisation'). The manufacturer must then draw up a Declaration of Conformity and affix the CE Marking.

The manufacturer will then:

- establish a procedure to review experience gained from products in the post-production phase and apply any corrective action;
- inform the Notified Body of:
  - any malfunction or deterioration in the characteristics and/or performance of a product;
  - any problems or changes being made to labelling or the Work Instructions;
  - any technical or medical reason connected with the characteristics or performance of a product leading to recall of products of the same type.

3.3.7.2 Notified Body

The Notified Body must complete a thorough examination (including tests) to check the conformity of the product with the requirements of the applicable Directive. This can be achieved either by examining and testing every product or (as is usually the case), examining and testing a small sample of the product (i.e. statistical examination). If using a statistical basis for this examination, the Notified Body will:

- require that the manufacturer presents his products in the form of homogeneous lots;
- perform an evaluation on a random sample taken from each lot;
- ensure that the sample meets the requirements of the Directive;
- determine whether the sample is acceptable or should be rejected.

If the sample is accepted, then the Notified Body will draw up a written Certificate of Conformity relating to the tests that have been completed and affix, or have affixed, an identification number for each approved product and draw up a written Certificate of Conformity. (The manufacturer may, if authorised by the Notified Body, affix the Notified Body's identification number during the manufacturing process.)

If the lot is rejected, the Notified Body shall ensure that appropriate measures are put in place to ensure that the product is not marketed.

3.3.7.3 Statistical Verification

For this type of verification, the manufacturer will present the manufactured products in the form of homogenous batches. The Notified Body will then take a random sample from each batch and this sample will be subjected to an individual test as defined in the relevant standard(s).

Statistical product control will be based on attributes, entailing a sampling system that ensures a limit quality corresponding to an acceptance probability of 5 per cent, with a non-conformity percentage of between 3 and 7 per cent. The sampling method will be established by the harmonised standards.

If the batch is accepted, the Notified Body will put an identification number on each product and draw up a written Certificate of Conformity relating to the tests carried out. All products in the batch may be put on the market except any in the sample which failed to conform.

If a batch is rejected, the Notified Body will take appropriate measures to prevent the batch from being placed on the market. In the event of frequent rejection of batches, the Notified Body may suspend the statistical verification.

3.3.7.4 Other Considerations

The manufacturer (or his authorised representative) must, for a period ending at least five years (varies with Directives) after the last product has been manufactured, keep at the disposal of the national authorities the following documentation:

- the Declaration of Conformity;
- his approved Quality Management System;
- details of any changes that have been made to the product following approval;
- details of the product's design, manufacture and performance;
- copies of all decisions and reports made by Notified Body.

3.3.8 Module G

Module G Unit Verification	Covers the design and production phases. Each individual product is examined by a Notified Body, which issues a certificate of conformity.	This consists of two parts: The **manufacturer ensures and declares** that the product(s) meet the requirements of the EC-Type Examination certificate(s) that has (have) been issued for it (them) in addition to the requirements of the applicable Directive. The **Notified Body examines and tests** the manufacturer's products for conformity with the relevant requirements of the appropriate Directive.

The product suppliers must be able to prove, through their documentation, that adequate design, production, and quality control procedures were in place to ensure a compliant product, that is, without defects.

3.3.8.1 The Manufacturer

The manufacturer is responsible for:

- affixing CE Marking to each product;
- drawing up a Declaration of Conformity.

3.3.8.2 Notified Body

The Notified Body must:

- complete a thorough examination (including tests) to check the conformity of the product with the requirements of the applicable Directive;
- affix its identification mark to the approved product;
- draw up a Certificate of Conformity concerning the tests that have been carried out.

3.3.9 Module H

Module H Full Quality Assurance	Covers the design and production phases. Derives from quality assurance standard EN ISO 9001, with the intervention of a notified body responsible for approving and controlling the quality system set up by the manufacturer.	**Manufacturers declare** that the products concerned satisfy the requirements of the applicable Directive, are in **conformity** with the EC-Type Examination certificate(s) that has (have) been issued for it (them) and that they operate a full **ISO 9001:1994** approved Quality Management System. The manufacturer then affixes CE Marking (accompanied by an identification symbol of the Notified Body responsible for the EC monitoring) to each product and draws up a written Declaration of Conformity.

To ensure compliance with this module the manufacturer must operate an approved quality system for production, final product inspection and testing in accordance with ISO 9001:1994. This quality system must ensure compliance of the products with the requirements of the Directive(s) that apply to them.

3.3.9.1 Quality System

The manufacturer must lodge an application for assessment of his quality system with a Notified Body of his choice, for the products concerned. This application must include:

- all relevant information for the product category envisaged;
- the documentation concerning the quality system.

All the elements, requirements and provisions that have been adopted by the manufacturer must be documented in the form of written policies, procedures and instructions. This documentation will form the basis of the manufacturer's Quality Management System (QMS) and will include quality programmes, plan, manuals and records.

In particular, a manufacturer's Quality Management System must contain a description of:

- the quality objectives and the organisational structure, responsibilities and powers of the management with regard to design and product quality;
- the technical design specifications (including standards);
- the design control and design verification techniques, processes and systematic actions that will be used;
- the manufacturing, quality control and quality assurance techniques, processes and systematic actions that will be used;
- the examinations and tests that will be carried out before, during and after manufacture, and the frequency with which they will be carried out;
- the quality records, such as inspection reports and test data, calibration data, qualification reports of the personnel concerned, etc.;
- the means to monitor the achievement of the required product quality and the effective operation of the quality system.

3.3.9.2 Notified Bodies

The Notified Body will assess the manufacturer's Quality Management System for conformance with ISO 9001:1994. At least one member of the auditing team must have experience of evaluating the product technology concerned and the evaluation procedure must include an inspection visit to the manufacturer's premises.

The Notified Body will also carry out periodic audits to make sure that the manufacturer maintains, applies and duly fulfils the obligations arising out of the approved quality system. The manufacturer must allow the Notified Body entrance (for inspection purposes) to the locations of manufacture, inspection and testing, and storage and must provide the Notified Body with all necessary information, in particular:

- the Quality Management System documentation;
- quality records for the design part – such as results, analyses, calculations, tests etc.;
- quality records of the manufacturing part – such as inspection reports, test and calibration data, qualification reports of the personnel, etc.

Under the terms of this module, the Notified Body may also pay unexpected and unannounced visits to the manufacturer to verify and validate that the quality system is functioning correctly.

The manufacturer must keep at the disposal of the national authorities for a period of 10 years (can vary with Directives) after the last of the product has been manufactured:

- the documentation associated with his QMS;
- details of any amendments, alterations or modifications to the QMS;
- previous reports and decisions made by the Notified Body.

Notified Bodies are required to provide other Notified Bodies with the relevant information concerning the quality system approvals that they have issued and withdrawn.

3.4 Choice of Module

The choice of which module shall be used largely depends on whether the manufacturer is involved in the design or the production of an industrial product. The differences are shown in Figure 3.3.

3.5 CE Marking

The essential objective of a conformity assessment procedure is to provide the public authorities with an assurance that products that have been placed on the market conform to the requirements expressed in the Directives – particularly those concerning the health and safety of users and consumers.

CE Marking symbolises conformity to all the obligations made on manufacturers for a product by the relevant Directive. Such conformity is

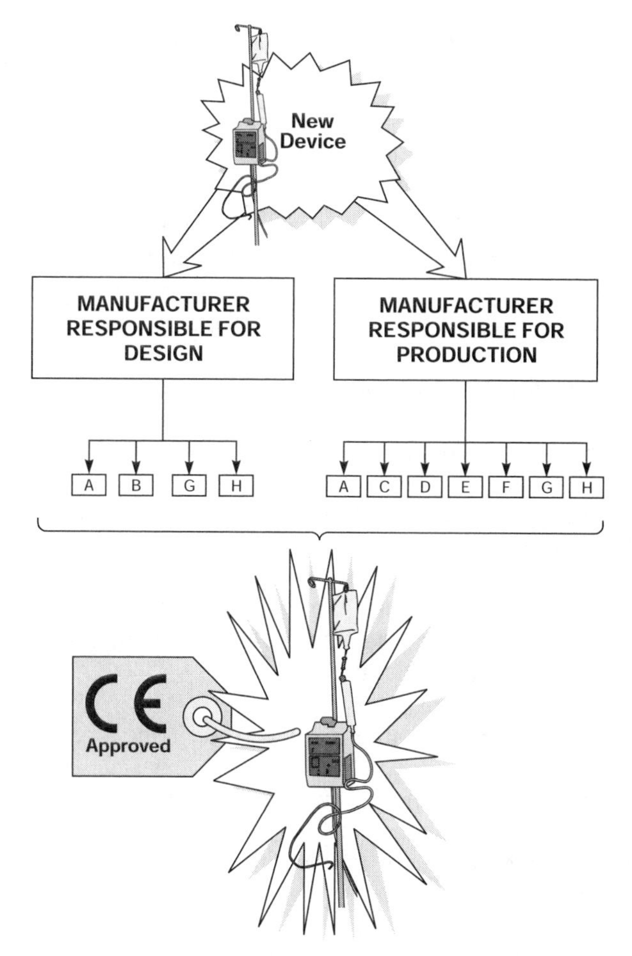

Figure 3.3 Design/production modules

not just limited to the Essential Requirements relating to safety, public health, consumer protection, etc. Certain Directives may also impose specific obligations not necessarily forming part of the Essential Requirements.

CE Marking is the **only** marking that may be used to certify that an industrial product conforms to the relevant Directive(s). To further emphasise this point, Member States have agreed that they will not mention (or introduce) any other form of Conformity Marking **other than CE Marking** into any of their national regulations.

If industrial products are subject to a number of Directives (which also provide for affixing the CE Marking), then the products concerned must indicate that they conform to the provisions of all of the appropriate Directives. When this happens, particulars of these Directives must be as published in the *OJEC* as well as being stated in the documents, notices or instructions accompanying the products and where appropriate, on the data plate.

3.5.1 General points regarding CE Marking

- CE Marking is officially referred to as such, and **not** as 'the CE Mark'.
- CE Marking is mandatory and must be affixed before any product covered by New Approach Directives, or other Directives providing for its affixing, is placed on the market and put into service, save where specific Directives require otherwise.
- Where industrial products are subject to several Directives, which all provide for the affixing of CE Marking, the Marking must indicate that the products are presumed to conform to the provisions of all these Directives.
- A product may not be CE marked, unless it is covered by a Directive providing for its affixing.
- The obligation to affix the CE Marking extends to all products within the scope of Directives providing for its affixing, and which are intended to be placed and/or put into service on the EU market for the first time. Thus, the CE Marking must be affixed:

  - to all new products, whether manufactured in the Member States or in third countries;
  - to used and second-hand products imported from third countries; and
  - to substantially modified products that are subject to Directives as new products.

- Directives may exclude the application of the CE Marking on certain products (such as medical devices) even when the Directive otherwise applies to the product. As a general rule, such products are subject to free circulation, if:

  - they are accompanied by a Declaration of Conformity (i.e. safety components referred to in the machinery Directive);
  - they are accompanied by a statement (i.e. custom-made medical products or products intended for clinical investigations);
  - they are accompanied by a Certificate of Conformity (i.e. components referred to in the Directive relating to potentially explosives atmospheres, which are intended to be incorporated into equipment or protective systems, and fittings referred to in the Directive relating to gas appliances);
  - the product bears the manufacturer's name and an indication of maximum capacity (i.e. instruments not subject to conformity assess-

ment according to the Directive relating to non-automatic weighing instruments); or
- the product is manufactured in accordance with sound engineering practice (i.e. certain vessels referred to in the Directive relating to simple pressure vessels).

- Industrial products for which CE Marking is a legal requirement (i.e. national legislation) do not require additional marking to indicate that the industrial product in question is legal.
- CE Marking must, when practicable and appropriate, be visible (i.e. at least 5 mm in height), legible and in an indelible form. If the CE Marking is reduced or enlarged the proportions given in the graduated drawing shown in Figure 3.5 must be used.
- CE Marking must be affixed to the product or to its data plate. Where this is not possible, it must be affixed to the packaging, if any, and to any accompanying documents. For medical equipment, it shall be shown on the product and/or on its sterile pack and associated instructions for use.
- If required, CE Marking may also be used by the manufacturer on his sales literature that accompanies the industrial product. Manufacturers are even allowed to use it in an advertisement. An identification number showing the Notified Body responsible for the product's assessment always accompanies the CE Marking.
- The continuing application of voluntary certification marks of quality (e.g. BSI Registered Firm), safety (e.g. BSI Safety Mark) or conformity (e.g. BSI Kitemark) criteria is unaffected by the introduction of the CE Marking regime.
- Voluntary certification marks provide **no** legal guarantee. Responsibility for the quality and safety of the industrial product remains with the manufacturer (or within the EU, the initial supplier). The effect of voluntary third-party certification is meant to provide additional assurance and provide a basis for increased confidence in the industrial product. ***Note:*** Such marks may only be affixed to the product, its packaging or the documentation accompanying the product on condition that the legibility and visibility of the CE Marking are not reduced. **Anybody** can offer a certification-marking scheme linked to the requirements of a national or European standard – or, for that matter, to any other specification they choose. Only market forces will determine the value of that mark!
- Products (other than products which are custom-made or intended for clinical investigations) that may be required to meet the Essential Requirements of the Directive must have CE Marking when they are placed on the market.
- CE Conformity Marking shall consist of the initials 'CE' taking the form shown in Figure 3.4.

Figure 3.4 CE Marking

3.5.2 Displaying the CE Marking

CE Marking replaces all mandatory conformity markings having the same meaning as CE Marking, and which might have existed in the national legislation of the Member States before harmonisation took place.

The affixing of legal marking (e.g. a protected trademark of a manufacturer or of a Notified Body), or of private certification marks additional to the CE Marking, is allowed to the extent that such markings or marks do **not** create confusion with the CE Marking, and do not reduce the legibility and visibility of the CE Marking.

Attaching or fixing CE Marking on an item indicates that the relevant Directive(s) associated with that particular industrial product have been complied with. Displaying CE Marking on an industrial product or its packaging is mandatory for most types of industrial product, but doing so implies that the goods comply with **all** relevant European Directives. Although the display and actual appearance of CE Marking may vary with a particular industrial product (e.g. if it is a very small item), at all times it must conform to the requirements detailed in 3.5.3 below.

3.5.3 Affixing CE Marking

Thus CE Marking affixed by the manufacturer (or his authorised representative in the EU) to industrial products 'symbolises the fact that the product conforms to all the Community total harmonisation provisions which apply to it and has been the subject of the appropriate conformity evaluation procedures'.

In **all** cases, CE Conformity Marking shall consist of the initials 'CE' taking the form shown in Figure 3.5.

If the CE Marking is reduced or enlarged the proportions given in Figure 3.5 **must** be respected. In addition, unless otherwise stated in a Directive, the height of the CE Marking shall be a minimum of 5 mm.

Any industrial product covered by the New Approach must bear the CE Marking and this must be affixed, visibly, legibly and indelibly, to the product or to its data plate. Where this is not possible, it can be affixed to the packaging, if any, and to any accompanying documents.

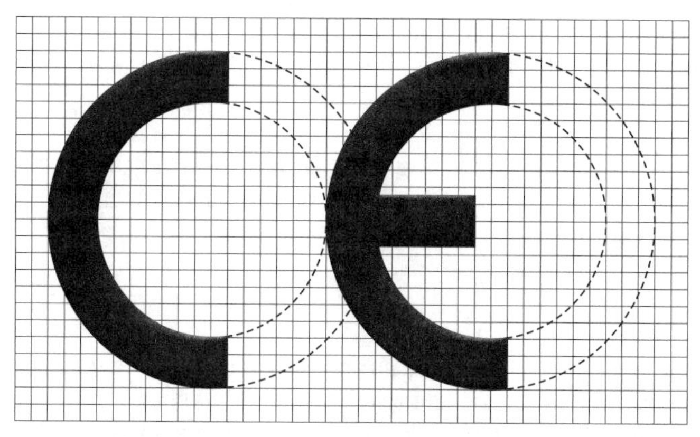

Figure 3.5 CE Marking – recognised format

This conformity is not just limited to the Essential Requirements found in each Directive (relating to safety, public health, consumer protection, etc.); certain Directives may also impose some additional (Directive-specific) obligations. Putting CE Marking onto an industrial product symbolises the fact that the product conforms to **all** of the EU total harmonisation provisions which apply to it and that the product has been the subject of the appropriate conformity evaluation procedures.

In principle, the CE Marking may not be affixed until the conformity assessment procedure has been completed, and it has been verified that the product complies with all the provisions of the relevant Directives. Usually this will be at the end of the production phase.

In some cases the CE Directive allows CE Marking to be affixed to the product's data plate, to the packaging or in certain cases (depends on the specific Harmonised Directive) for it to be included in the accompanying documentation. Assuming that the CE Marking is on a data plate that is not affixed to the product until **after** the final inspection, this should not cause any problems. If, however, the CE Marking forms an inseparable part of the product then the Marking may be affixed at any other stage of the production phase, **provided** that the manufacturer or the Notified Body verifies the conformity of the product in the production control phase.

The CE Marking on the product may be neither omitted nor be moved to the packaging or accompanying documents on purely aesthetic grounds. Only where the general rule regarding the affixing of the CE Marking cannot be followed, may the CE Marking, **as an exception**, be moved from the product or its data plate. This can be justified because:

- the affixing to the product is impossible (e.g. on certain types of explosives);
- it is not possible under reasonable technical conditions;

- the minimum dimensions cannot be respected;
- it cannot be ensured that the CE Marking is visibly, legibly and indelibly affixed.

In such cases, the CE Marking has to be affixed to the packaging, if it exists, and to the accompanying document (always assuming that the Directive concerned provides such documents).

Notes

The actual wording of this requirement differs with Directives. According to the Directives relating to simple pressure vessels, machinery, non-automatic weighing instruments, gas appliances, telecommunication terminal equipment, hot-water boilers, recreational craft, lifts, refrigeration appliances and pressure equipment, the CE Marking **always** has to be affixed to the product. The Directive relating to potentially explosive atmospheres requires that the CE Marking **must** be affixed to the product.

However, it has been agreed that the general principle of moving the CE Marking from the product or its data plate to, for instance, the packaging, is also applicable for these products under similar circumstances provided that it is included in the applicable Directives as an option.

The requirement for **visibility** means that the CE Marking must be easily accessible for market surveillance authorities. By the same token, it must also be visible for distributors, users and consumers. It may therefore, for example, be affixed on the back (e.g. television set) or underside of a product (e.g. telephone/facsimile machines). For this reason (i.e. legibility) a minimum height of 5 mm of the CE Marking was introduced.

According to the Directives relating to Active Implantable Medical Products, Medical Devices, and Potentially Explosives Atmospheres, the minimum dimension of the CE Marking may be waived, but only for small products.

The requirement for **indelibility** means that the CE Marking must not be removed from the product without leaving traces noticeable under normal circumstances. However, this does not mean that the CE Marking must form an integral part of the product.

Depending on the conformity assessment procedure(s) that are applied, a Notified Body may be involved in the design phase, the production phase or in both phases. However, only if the Notified Body is involved in the production phase will its identification number follow the CE Marking. For example, the identification number of the Notified Body involved in Module B conformity assessment never follows the CE Marking.

Sometimes several Notified Bodies are involved in the production phase – especially where more than one Directive is applicable. In these situations several identification numbers may follow the CE Marking. Accordingly, the CE Marking will appear on products either:

- without an identification number, which means that a Notified Body did not intervene in the production phase (e.g. module A, modules Aa and C (where the Notified Body only intervened during the design phase) and the combination of modules B and C);
- with an identification number, which means that the Notified Body assumes the responsibility for:
  - the examinations and tests carried out to assess the conformity of the product in the production control phase (modules F, Fbis and G);
  - for the tests on specific aspects of the product (modules Aa1 and Cbis1 where the Notified Body intervened during the manufacturing phase);
  - product checks (modules Aa2 and Cbis2); or
  - the assessment of production, product quality assurance or full quality assurance (modules D, E and H).

The CE Marking and the identification number of the Notified Body do not necessarily have to be affixed within the EU. These can be affixed in a third country, if the product, for example, is manufactured there and the Notified Body carried out conformity assessment in accordance with the Directive in that country. The CE Marking and the identification number can also be affixed separately, as long as the Marking and number remain combined.

The CE Marking is affixed at the end of the production control phase and has to be followed by the identification number of the Notified Body. These numbers are assigned by the Commission as part of the notification procedure and are published in the *OJEC*. If a Notified Body is notified under several Directives, then they shall be assigned the same number. Thus:

- the CE Marking must be affixed by the manufacturer (or his agent), or in certain cases it can be affixed by the person responsible for placing the product on the Community Market;
- the identification number of the Notified Body must be affixed either by the Body itself, or by the manufacturer.

The CE Marking consists exclusively of the letters 'CE' followed by the identification number(s) of any Notified Body involved in the production phase. This can then be followed by a pictogram (or any other mark) indicating, for example, the category of use – but any other marking which is likely to deceive third parties as to the meaning and form of the CE Marking, is prohibited.

Thus a product may bear different marks (for example marks indicating conformity to national or European standards or with traditional optional Directives), provided such marks are not likely to cause confusion with the CE Marking.

3.5.3.1 Equipment for Connection to Public Telephone Network

For equipment that has been approved for connection to a Public Telecommunications Network (PTN), there are actually three marks:

- The first is CE Marking.
- The second mark is a four digit code which symbolises the body who approved the equipment, for example the number for BABT (the British Approvals Board for Telecommunications) is 0168.
- The third mark looks like crossed hockey sticks and indicates that the equipment is suitable or not suitable for connection to the Public Telecommunications Network (PTN).

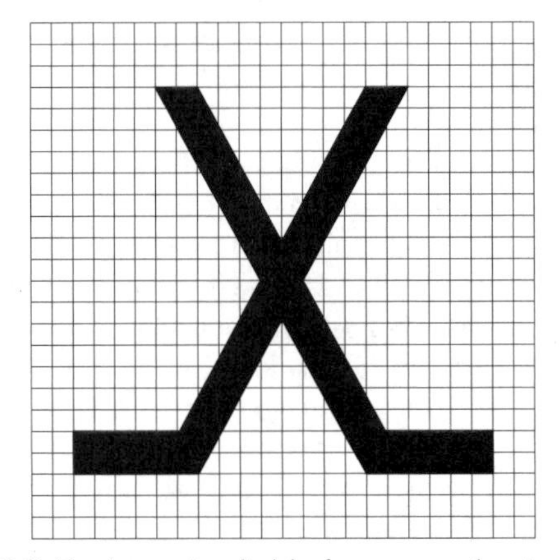

Figure 3.6 Equipment suitable for connection to the PTN

Figure 3.7 Equipment not suitable for connection to the PTN

3.5.3.2 Marine and High Speed Rail

The Directives relating to marine equipment and high-speed rail equipment follow the principles of Global Approach, but do not require CE Marking. Instead the Directive on marine equipment requires a special conformity mark.

3.5.3.3 Hot-water Boilers and Refrigeration Appliances

The Directives relating to hot-water boilers and refrigeration appliances follow the principles of Global Approach – but not the principles of New Approach – and provide for the affixing of the CE Marking.

3.5.3.4 Precious Metals

A different Marking instead of the CE Marking is foreseen in the draft Directive on articles of precious metal.

3.5.3.5 Potentially Explosive Atmospheres

The explosion protection symbol (see Figure 3.8) is required for equipment and protective systems intended for use in potentially explosive atmospheres. This symbol is normally followed (i.e. associated with) the equipment group and category.

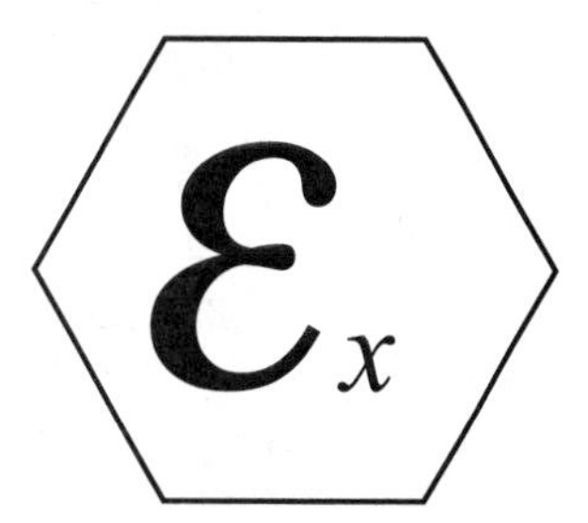

Figure 3.8 Explosion protection symbol

3.5.3.6 Others

The Directives relating to construction products, machinery, gas appliances, active implantable medical products, potentially explosive atmospheres, medical products, and personal protective equipment require additional labelling, for instance, to identify the manufacturer and the product, or to point out the safe use of the product.

3.5.4 Wrongly Affixed CE Marking

Member States are required (under national law) to prevent abuse of the CE Marking and to avoid any possibility of confusion with its use. If a Member State finds out that CE Marking has been put onto an unauthorised product, then the Member State shall make the manufacturer:

- either make the product comply;
- restrict (or prohibit) the placing on the market of the product in question;
- or ensure that the product is withdrawn from the market.

It is prohibited to affix marks or inscriptions that are likely to mislead third parties with regard to the meaning and/or the graphics of CE Marking. Where CE Marking has been fixed to a non-compliant industrial product, the manufacturer is obliged to end the infringement under the conditions imposed by the Competent Authority.

3.5.5 Other Considerations

Specific Directives may allow CE Marking to be affixed to the packaging or the accompanying documentation, instead of to the product itself.

The Declaration of Conformity (or the Certificate of Conformity depending on the Directive concerned) must cover either individual or several products and shall either accompany the product(s) or be retained by the manufacturer.

The development of the online (e.g. e-mail) availability of certificates and other documents issued by Notified Bodies is currently being discussed.

3.5.6 Withdrawal of Product

If a Member State finds out that the CE Marking has been put onto an unauthorised product, then the manufacturer is obliged to make the product comply and to end the infringement under conditions imposed by the Member State. Where non-compliance continues, the Member State must take all appropriate measures to restrict (or prohibit) the placing on the market of the product in question or ensure that it is withdrawn from the market in accordance with the procedures laid down in the safeguard clauses contained in the relevant Directive.

Prior to the New Approach Directives, the manufacturer had only to show that reasonable measures had been taken to ensure a safe product. In 1985 the Product Liability Directive was issued, changing the old focus of 'negligence' to that of 'strict liability', on the manufacturer's or supplier's part, thereby shifting the onus better to protect the consumer and placing a greater burden on the product supplier.

3.6 Industrial Product Type Conformity

It is essential that any manufacturer or importer who intends to sell industrial products within the EU is familiar with the requirements imposed not just by the CE Conformity Marking Directive but by the various other Directives which may also apply to their industrial products. This is an important part of the Single Market initiative which is aimed at removing barriers to trade between Member States of the EU and which also extends to European Free Trade Association (EFTA) countries and to other countries (e.g. the USA) with which the EU has negotiated protocols and agreements. The main point is that if your industrial product meets the requirements for CE Marking it can be sold anywhere in the European Economic Area and is (virtually) acceptable world-wide, **without** having to meet further technical requirements.

It cannot be overstated that **only** industrial products which comply with all relevant Directives **may** carry CE Marking!

3.6.1 Industrial products which do not need to be CE marked

For the purpose of the CE Directive, it has been agreed that the following products do not need to be CE marked:

- products intended for clinical investigation being made available to medical practitioners or authorised persons;
- custom-made products being placed on the market or being put into service.

3.7 Competent Authority

A Competent Authority (sometimes referred to as a 'National Authority') is a regulatory Body within a member state that is charged with ensuring that the provisions of Directives are fully implemented. For industrial products in the UK, the Competent Authority is the Secretary of State for Health acting through the Industrial Products Agency.

The Competent Authority is also responsible for ensuring that all industrial products placed on the market or put into service meet the Essential Requirements laid down in the Directive.

3.7.1 General responsibilities

The general responsibilities of the Competent Authority include:

- negotiating the Directives on behalf of the UK Government;
- putting in place the necessary Statutory Instruments to give effect to the Directives;

- liaising and consulting with interested parties within the Department of Health, other Government Departments, the National Health Service and other bodies, including the UK healthcare industry;
- participating in EC Working Parties and other Groups;
- publicising the provisions of the Directives and issuing appropriate guidance to product users and manufacturers in the UK;
- advising on the CE Conformity Marking Directives (e.g. interpretation of the Articles) conformity assessment routes and the classification of products; and
- monitoring the effectiveness of the Regulations and their burdens on business.

3.7.2 Specific tasks

More specific tasks of the Competent Authority include:

- enforcing the Regulations using powers under the Consumer Protection Act;
- liaising with the EU Commission if it considers that harmonised standards do not meet the Essential Requirements;
- membership of the Standing Committees set up under the Directives to provide opinions, advice and guidance to Member States;
- undertaking appropriate measures to withdraw unsafe products from, or prohibit or restrict them from being placed on, the market;
- resolving disputes between (for example) manufacturers and Notified Bodies;
- ensuring that adverse incidents are reported within the appropriate timescale and are recorded and evaluated centrally;
- handling all applications for clinical investigations, including arrangements for evaluation by expert assessors; and,
- designating Notified Bodies within the UK to carry out conformity assessment procedures.

The Competent Authority must ensure, except for custom-made products and those intended for clinical investigation, that:

- **all** products carrying the CE Marking meet the Essential Requirements of the relevant Directives;
- products **not** meeting the Essential Requirements of the relevant Directives are **not** allowed to carry the CE Marking;
- except for products marketed under transitional arrangements, **only** products carrying the CE Marking are placed on the market.

3.8 Notified Bodies

Notification	(in the context of CE Marking) is the formal recognition of a testing and/or certification body by the European Commission (EC).
Certificate	is a visible form of attestation through an independent and impartial third party.
Certification	means that a product or service conforms to certain technical rules contained in standards and regulations.
Accreditation	means the recognition of the competence of a laboratory, certification or inspection body by independent accreditors (e.g. German Accreditation Council (DAR), NAMAS (UK) and RNE (France) etc.

A Notified Body (e.g. BSI Industrial Product Certification or Amtec Certification Services Ltd) is a certified organisation that the Competent Authority designates to carry out one or more conformity assessment procedures, when a third party is required. Put simply, the Notified Bodies are authorised to approve the procedures used by a manufacturer to produce an industrial product.

Under certain New Approach Directives this Body is not called a Notified Body but instead an 'Inspection Body' (e.g. for the simple pressure vessels and construction products Directives) or a 'Testing Laboratory' and/or a 'Certification Body' (e.g. for the Construction Products Directive) or an 'Approved Body' (e.g. for the Toys Directive). However, the same principles for Notified Bodies are applicable to these bodies

Member States must verify the competence of the bodies seeking notification. This shall be based on the criteria laid down in the applicable Directive in conjunction with Essential Requirements and the conformity assessment procedure in question. In general, the competence criteria set out in the Directives cover:

- availability of personnel and equipment;
- independence and impartiality in relation to those directly or indirectly concerned with the product (e.g. designer, manufacturer, manufacturer's authorised representative, supplier, assembler, installer, user);
- technical competence of personnel that is relevant to the products and conformity assessment procedure in question;
- maintenance of professional secrecy;
- subscription to civil liability insurance, unless that liability is covered by the state under national law.

	Certification Bodies	*Testing Laboratories*	*Inspection Bodies*
Criteria for Accreditation Bodies	EN 45010	EN 45003	EN 45010
Accreditation and assessment criteria	EN 45010	EN 45003	EN 45010
Operational criteria	EN 45011 EN 45012 EN 45013	EN 45001	EN 45004

Table 3.4 The EN 45000 series of standards relevant for Notified Bodies

The assessment of the body seeking notification will determine if the body fulfils the requirements. Although not compulsory, accreditation is normally completed in accordance with the EN 45000 series of standards. This series covers all the different types of Conformity Assessment Bodies, i.e. Certification Bodies, Testing Laboratories, Inspection Bodies and Accreditation Bodies. Indeed, it is irrelevant whether the body calls itself a Laboratory, a Certification Body or an Inspection Body as long as it carries out the tasks in the conformity assessment procedure and has the technical ability to do so in an independent and impartial way.

The EN 45000 standards consist, in general terms, of one part dealing with the organisation and management of the body, and another part dealing with the technical requirements related to the operation of the body. The standards must be seen as an integral whole, since both parts are required to ensure the reliability and capability of the operation of the Conformity Assessment Bodies. For the assessment of competence of bodies seeking notification, the essential standards are:

3.9 Essential Standards for Notified Bodies

Essential Standards for Notified Bodies are shown in Table 3.5.

3.10 Relevant Standards of the EN 45000 Series for each module

The relationship of these standards to the various modules is shown in Table 3.6.

Member States are responsible for ensuring that Notified Bodies maintain their competence at all times, are capable of carrying out the

EN 45001	General requirements a laboratory must meet if it is to be recognised to carry out testing and/or calibration.
EN 45004	Specifies the general criteria for the competence of bodies performing inspection. Inspection involves examination of a product design, product, service, process or plant and determination of their conformity with specific requirements or, on the basis of professional judgement, general requirements.
EN 45011	Specifies the general requirements that a third party operating a product certification system must meet. Product certification entails assurance that a product conforms to specified requirements such as standards, regulations, specifications or other normative documents.
EN 45012	Specifies the general requirements that a third party operating quality system certification must meet. Quality system certification involves the assessment, determination of conformity against quality system standard and within a certain scope of activity and surveillance of the supplier's quality system.

Table 3.5 Essential standards for Notified Bodies

Module	*EN 45001*	*EN 45004*	*EN 45011*	*EN 45012*
A	Yes	Yes	Yes	
B		Yes	Yes	
C	Yes	Yes	Yes	
D				Yes
E				Yes
F	Yes	Yes	Yes	
G		Yes	Yes	
H		Yes	Yes	Yes

Table 3.6 Relevant standards of the EN 45000 series for each module

work for which they are notified and can complete this in accordance with the varied requirements of the relevant Directives. For example:

- The Directives relating to safety of toys, construction products, electromagnetic compatibility, personal protective equipment, non-automatic weighing instruments and gas appliances set up clearly an obligation for the Member States to verify or have verified periodically the fulfilment of the conditions concerning the availability, competence and integrity of personnel, and the necessary means and equipment.
- The Directive on marine equipment requires that Member States organise an audit every two years to ensure that each Notified Body continues to comply with the criteria.

3.11 Notified bodies' tasks under each module

After a successful assessment and recognition by the Competent Authority, the State government notifies the Commission of a potential Accredited Body. The Accredited Bodies are then listed in the *Official Journal of the European Community* (*OJEC*). As of 1999, it has been estimated that currently there are over 10,000 testing laboratories and 1,000 certification bodies in Europe of varying capacity, legal status and reputation as Bodies Notified to the Commission and to the Member States, and these are commonly referred to as Notified Bodies (safety), Competent Bodies (EMC), or referred to in general terms as third parties.

The published list of the Notified Bodies (together with the identification numbers that have been allocated to them and the tasks for which they have been notified) in the current list of UK Notified Bodies can be obtained from the Industrial Products Agency web-site at

http://www.medical-products.gov.uk.

When a product becomes suspect, the national enforcement authority will typically use the EN or equivalent national standards (usually with the help of a Notified Body) to evaluate the product's conformity or lack thereof.

3.11.1 Role of the Notified Body

The role of the Notified Body is to:

- ensure conformity, build consumer confidence, and protect public interests;
- provide facilities for conformity assessment on the conditions set out in the Directives;
- issue test reports and certificates on conformity;

Module	*Notified Body*
A	• No requirement
Aa1	• Supervises the tests carried out by the manufacturer • Supervises the affixing of its identification number, where it was involved in conformity assessment during the production stage • Keeps a record of relevant information • Communicates to the other Notified Bodies relevant information (on request)
Aa2	• Carries out or has carried out product checks at random intervals, and for this purpose takes samples of final products • Supervises the affixing of its identification number • Keeps a record of relevant information • Communicates to the other Notified Bodies relevant information (on request)
B	• Ascertains, by performing or having performed examinations and tests, that the specimen(s) meet(s) the applicable provisions and is manufactured in accordance with the technical documentation • Issues an EC-Type Examination certificate • Keeps a copy of the certificate and a record of other relevant technical information • Communicates to the other Notified Bodies the relevant information concerning the EC-Type Examination certificates (on request)
C	• No requirement
D	• Assesses the quality system to determine whether it satisfies the applicable requirements, and accordingly takes a decision
E	• Supervises the affixing of its identification number • Carries out surveillance of the manufacturer by means of periodical and unexpected visits • Keeps a record of relevant technical information • Communicates to the other Notified Bodies the relevant information concerning the quality system approvals issued and withdrawn (on request)
F	• Carries out the appropriate examinations and tests in order to check the conformity of the product with the applicable requirements either by examination and testing of every product, or by examination and testing of products on a statistical basis • Supervises the affixing of its identification number • Draws up a Certificate of Conformity relating to the tests carried out • If a lot is rejected, takes appropriate measures to prevent the putting on the market of that lot • Keeps a record of relevant technical information • Communicates to the other Notified Bodies relevant information (on request)
G	• Examines the individual product, and carries out the appropriate tests to ensure its conformity with the relevant requirements • Supervises the affixing of its identification number • Keeps a record of relevant information • Draws up a Certificate of Conformity concerning the tests carried out • Communicates to the other Notified Bodies relevant information (on request)
Hbis	• In addition to responsibilities as in module D: • Examines the application • Issue an EC design examination certificate, if the design meets the applicable provisions • Keeps a record of the EC design examination certificates and the EC design approvals • Communicates to the other Notified Bodies of relevant information concerning the EC design examination certificates and the EC design approvals (on request)

Table 3.7 Notified Bodies' tasks under each module

- provide guidance on the Essential Requirements or other provisions of the Directives;
- provide technical interpretation and applicability of standards;
- be available to be called upon to assess manufacturer's technical file and documentation;
- determine alternatives when no standards exist or other criteria used for conformity to ERs; and
- test sample products for conformity and award 'Approval Marks', with ongoing surveillance of production, to visibly demonstrate product quality and increase marketing advantage.

3.11.2 Guidelines for Notified Bodies

- Notified Bodies are listed in the *Official Journal of the European Community*;
- Member states should accept test reports and/or certificates of all Notified Bodies;
- Testing may be subcontracted to other laboratories, under the direct supervision of the body; and
- Notified Bodies should reside in one of the EU member states.

3.11.3 Tasks of a Notified Body

The tasks of a Notified Body will vary depending on what conformity assessment route a manufacturer has chosen. For example:

- Where a manufacturer has chosen to follow the **full quality assurance route** (i.e. Module H) the Notified Body will carry out a complete assessment of the manufacturer's quality assurance system. If the manufacturer is required by a Directive to produce a design dossier the Notified Body will examine and evaluate it.
- Where a manufacturer has chosen to follow the **type-examination route** (i.e. Module B), the Notified Body may use harmonised European standards to assess whether the Essential Requirements have been met. Where there are no harmonised standards, the Notified Body (possibly in conjunction with the manufacturer) may devise their own means of assessing compliance with the Essential Requirements, possibly taking into account national and/or other standards if these appear to be adequate for this purpose.
- The Notified Body must apply the procedures described in the Directive for statistical verification, industrial production quality system audit or industrial product quality audit. The Notified Body carrying out this assessment can be different to that used for the type-examination assessment.

3.11.4 Criteria to be met for the Designation of Notified Bodies

New Approach Directives require that the Notified Body, its Director and the assessment and verification staff shall not be the designer, manufacturer, supplier, installer or user of the products which they inspect, nor the authorised representative of any of these persons.

The Notified Body and its staff must carry out the assessment and verification operations with the 'highest degree of professional integrity and the requisite competence in the field of industrial products'. They must be free from all pressures and inducements (particularly financial!) which might influence their judgement or the results of the inspection – especially from persons or groups of persons with an interest in the results of the verifications.

They shall meet legally binding criteria set out in the annexes to the Directives.

3.11.5 Requirements of a Notified Body

The Notified Body must have:

- sound vocational training;
- satisfactory knowledge of the rules on the inspections;
- adequate experience of such inspections;
- the ability required to draw up the certificates, records and reports.

Other requirements:

- the impartiality of the Notified Body must be guaranteed;
- the Notified Body must take out third party civil liability insurance;
- the staff of the Notified Body are bound to observe professional secrecy.

3.11.6 Specific requirements associated with CE Marking

Member States need to ensure that the Notified Bodies have the technical qualifications relevant to the appropriate Directive and that the latter keep their competent national authorities informed of their capability in this respect.

Notified Bodies will be encouraged to apply the modules without unnecessary burden to the 'economic operators'. In this context, the technical documentation that a manufacturer has to provide to the Notified Bodies will be limited to those that are required solely for the purpose of assessment of conformity.

Whenever a Directive provides the manufacturer with the possibility of using modules based on quality assurance techniques, the manufacturer must also be able to have recourse to a combination of modules not using quality assurance, and *vice versa*.

Notified Bodies who can prove their conformity with the EN 45000 series by submitting an accreditation certificate or other documentary evidence are presumed to conform to the requirements of the Directives.

A list of Notified Bodies is published by the Commission in the *Official Journal of the European Community* and is constantly updated.

3.11.7 Subcontracting part of the Notified Bodies work

Subcontracting part of the Notified Body's workload is permissible and shall be subject to:

- the competence of the subcontractor;
- conformity with the EN 45000 series of standards;
- the capability of the Notified Body;
- effective monitoring of such compliance;
- the ability of the Notified Body to exercise effective responsibility for the work carried out under subcontract.

4 REQUIREMENTS OF THE VARIOUS DIRECTIVES AFFECTED BY CE MARKING

4.1 Content of the Directives

The Directives are made up of a number of sections which, according to the particular Directive, describe:

4.1.1 Scope

- Objectives;
- Applications;
- Range of products;
- Nature of hazards intended to avert;
- Products excluded from the Directive.

4.1.2 General Clause for Placing Goods on the Market

- Safety of persons, domestic animals and/or property;
- Criteria for proper installation;
- Foreseeable use;
- Criteria for transportation.

4.1.3 Essential Requirements

- Legally binding obligations that include everything necessary for the protection of the public interest, i.e.:
  - safety of products;
  - protection of workers;
  - protection of consumers;
  - health;
  - environmental protection.

4.1.4 Free Movement of Goods

- Agreements and Requirements;
- Obligation of all Member States.

4.1.5 Proof of Conformity

- Conformity with harmonised standards;
- Conformity where a product does not, or only partly, complies with a harmonised standard;
- Conformity where no harmonised standard exists.

4.1.6 Management of the Standards

- CEN, CENELEC and ETSI harmonised standards;
- Publication in the *OJEC*;
- Publication of national standards.

4.1.7 Safeguard Clause

- Invoked if a product with CE Marking fails to comply with the safety requirements or the product is non-conforming.

4.1.8 Standing Committee

- Rules and details.

4.1.9 Means of Attestation of Conformity

- Chosen from the modules contained in Annexes to each Directive.
- Rules for affixing CE Marking concerning:
  - design;
  - manufacture;
  - placing on the market;
  - entry into service;
  - use.

4.2 Conformance

Unfortunately the technical and administrative requirements for CE Marking vary between Directives, within Directives and between different industrial product categories. For example, under the Construction Industrial Products Directive, CE Marking confirms compliance with a bespoke (i.e. 'mandated' or 'harmonised') European Standard or European Technical Approval. Under the Medical Devices Directive, CE

Marking needs some form of third-party (i.e. Notified Body) involvement in assessing the industrial product and/or the industrial production facility. For the vast majority of industrial products which require CE Marking, however, little more than a supplier's declaration that his 'industrial product meets the applicable legal requirements' of the relevant Directive is required.

4.2.1 CE Marking and Inscriptions

In all cases the following shall apply:

- the CE conformity Marking shall consist of the initials 'CE' taking the form shown in Figure 4.1;
- if the CE Marking is reduced or enlarged the proportions given in Figure 4.1 must be respected;
- the various components of the CE Marking must have substantially the same vertical dimension, which may not be less than 5 mm.

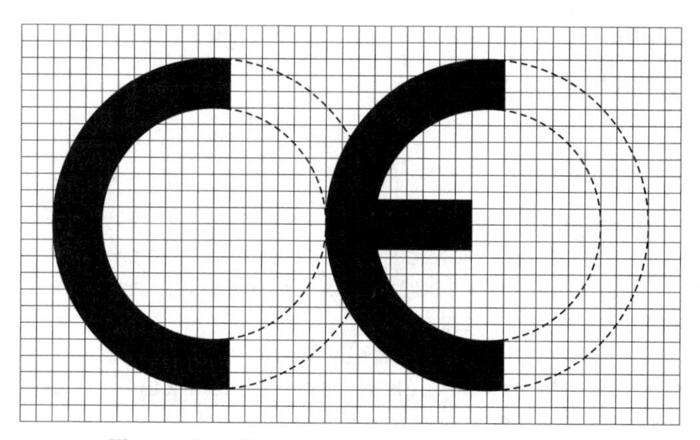

Figure 4.1 CE Marking – recognised format

Notes

Additions or amendments to the following rules are shown in the 'CE Marking' sub-section of the following descriptions of the Directives.

The requirements of the individual Directives that are affected by the CE Marking Directive are as follows:

4.3 Principal Directives

4.3.1 Low Voltage Equipment (73/23/EEC)

Directive:	**73/23/EEC** **Dated: 19 Feb 73**
Short name	Low Voltage Directive – LVD
Description	Safety of all electrical equipment operating between 50 and 1000 V
Amendments	93/68/EEC (CE Marking)
Repealed Directives	Nil
Member States Implementation Date	21 Aug 74 Implemented in the UK by The Electrical Equipment (Safety) Regulations 1994 (SI 1994/3260) which replaced the 1989 Regulations
Transitional Period	Until 31 Dec 96
Date of entry into force	1 Jan 97

4.3.1.1 Introduction

The Low Voltage Directive (LVD) is an electrical safety Directive that applies to all electrical products and equipment which are designed to operate in the ranges 50–1000 Vac or 75–1500 Vdc and whose hazards are primarily of an electrical nature.

The LVD's Essential Requirements have been in effect since 1973 and mandate conformity for all safety aspects of electrical products, including those of a more mechanical origin.

4.3.1.2 Structure

The Directive includes the Annexes listed in Table 4.1.

Annex	*Title*
I	Principal elements of the safety objectives for electrical equipment designed for use within certain voltage limits
II	Equipment and phenomena outside the scope of the Directive

Table 4.1 Annexes to Directive 73/23/EEC

4.3.1.3 Objectives

To harmonise the laws of Member States relating to low voltage electrical equipment that is designed to operate with a voltage rating of 50–1000 V (for alternating current) and 75–1500 V (for direct current).

4.3.1.4 Exclusions

The Directive applies to all low voltage equipment within the voltage ranges specified above, with the exception of:

- electrical equipment for use in an explosive atmosphere;
- electrical equipment for radiology and medical purposes;
- electrical fence controllers;
- electrical parts for goods and passenger lifts;
- electricity meters;
- plugs and socket outlets for domestic use;
- radio electrical interference;
- specialised electrical equipment for use on ships, aircraft and railways.

In addition, this Directive does not apply to electrical equipment intended for export to third countries.

4.3.1.5 Essential Requirements

Article 2 of the Directive requires that:

> electrical equipment may be placed on the market only if, having been constructed in accordance with good engineering practice in safety matters in force in the Community, it does not endanger the safety of persons, domestic animals or property when properly installed and maintained and used in applications for which it was made.

Also,

> 'Member States must ensure that stricter safety requirements than those laid down in Article 2 of the Directive are **not** imposed by electrical supply bodies for connection to the grid, or for the supply of electricity to users of electrical equipment.'

4.3.1.6 Principal Elements of the Safety Objectives

4.3.1.6.1 General Conditions

- The essential characteristics shall be marked on the equipment, or an accompanying notice.
- The manufacturers or brand name or trademark should be clearly printed on the electrical equipment or the packaging.

- The electrical equipment, together with its component parts, should be made in such a way as to ensure that it can be safely and properly assembled and connected.
- The electrical equipment should be designed and manufactured so as to ensure protection against the hazards (set out in paragraphs 2 and 3 of Annex 1 to the Directive).

4.3.1.6.2 Protection Against Hazards Arising from the Electrical Equipment

Observance of the general conditions (see above) shall ensure that:

- persons and domestic animals are adequately protected against danger of physical injury or other harm;
- temperatures, arcs or radiation which would cause a danger, are not produced;
- insulation must be suitable for foreseeable conditions.

4.3.1.6.3 Protection Against Hazards which may be Caused by External Influences on the Electrical Influence

Observance of the general conditions (see above) shall ensure that electrical equipment:

- meets the expected mechanical requirements;
- is resistant to non-mechanical influences in expected environmental conditions;
- does not endanger persons, domestic animals and property in foreseeable conditions of overload.

4.3.1.7 Proof of Conformity

Industrial products are presumed to conform to the safety objectives of the LVD where the equipment has been manufactured in accordance with harmonised technical standards.

Alternatively, the manufacturer may construct a product that is in conformity with the Essential Requirements and the safety objectives of

Title	*Yes/No*
CE Marking	Yes
EC Declaration of Conformity	Yes
Technical File	Yes

Table 4.2 Proof of Conformity (73/23/EEC)

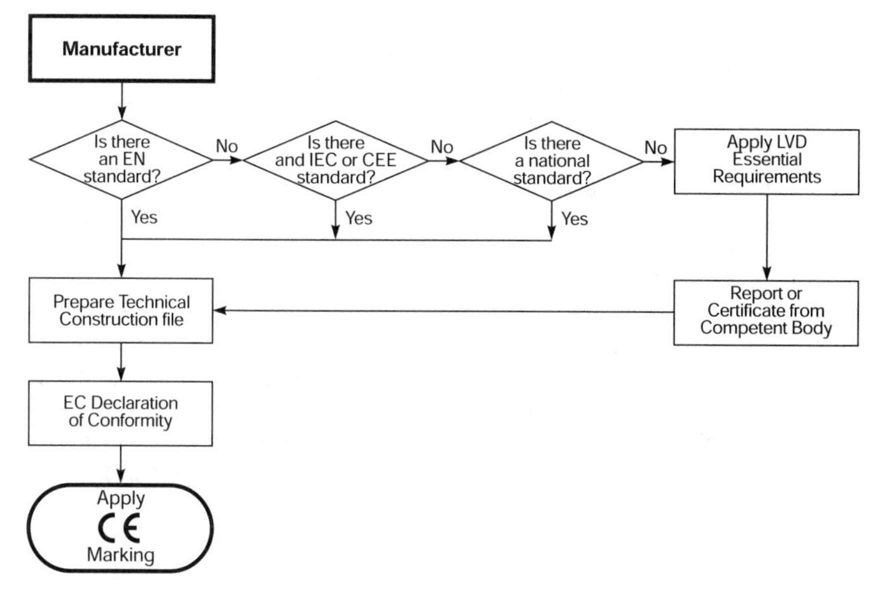

Figure 4.2 Conformity Procedure (73/23/EEC)

the Directive – without applying harmonised, international, or national standards. In such cases, the manufacturer must provide a description (in the technical documentation) of the solutions that have been adopted to satisfy the safety aspects of the Directive.

In order to certify the conformity of the electrical equipment with the safety provisions of the Directive and the CE Marking, the manufacturer must draw up a technical construction file covering the design, manufacture and operation of the electrical equipment, establish an EC Declaration of Conformity and affix the CE Marking.

Note: According to other reference documents from the Commission, the placing on the market refers to each individual product and not to a type or model of product.

4.3.1.8 Other Relevant Information

The Low Voltage (LVD) Directive (72/23/EEC) has now been adopted as European Standard UTEC 00–105 concerning electrical equipment installation.

4.3.1.9 Harmonised Standards

If the relevant harmonised standards are unavailable (e.g. they have not yet been drawn up and published), Member States are required to conform to the safety provisions of the International Commission on the rules for the Approval of Electrical Equipment (CEE) or of the International Electrotechnical Commission (IEC) procedure.

4.3.1.9.1 Publications

Application Guide (available from the Commission) 'Guidelines on the application of CD 73/23/EEC'.

There are also two Memoranda published by CENELEC that assist further understanding of the Directive:

- Memorandum No 3, 'On the implementation of the EEC Low Voltage Directive of 19 February 1973 with respect to reports, certificates and manufacturer's Declaration of Conformity'. November 1975, Amended March 1985 and March 1987.
- Memorandum No 6, 'Adoption of new standards as a basis for the certification of products in the CENELEC Member Countries'. October 1989.

4.3.1.9.2 CE Marking

As a result of the changes made in the CE Amending Directive (93/68/EEC), the procedures whereby a product carried a presumption of conformity if it was approved by a certification body has now been replaced by a new requirement for equipment to have the CE Marking after the self-certification procedure.

All these new measures became mandatory from 1 January 1997 although manufacturers could have chosen to comply with either the existing rules for CE Marking or the new rules until that date.

Member States were required to take the measures necessary to ensure the implementation of this amending Directive by 1 July 1994.

4.3.2 ElectroMagnetic Compatibility (89/336/EEC)

Directive:	**89/336/EEC Dated: 3 May 1989**
Short name	EMC
Description	Noise immunity and emissions performance of electrical equipment
Amendments	92/31/EEC (defining the transitional period) 91/263/EEC (TTE) 93/68/EEC (CE Marking) 93/97/EEC (satellite earth station equipment) 98/13/EC (TTE/SES)
Repealed Directives	76/889/EEC 76/890/EEC
Member States Implementation Date	1 Jul 92 Implemented in the UK by The Electromagnetic Compatibility Regulations 1992 (SI 1992/2372) and The Electromagnetic Compatibility (Amendment) Regulations 1994 (SI 1994/3080)
Transitional Period	Until 31 Dec 95
Date of entry into force	1 Jan 96

4.3.2.1 Introduction

ElectroMagnetic Compatibility (EMC) is the ability of the apparatus to operate satisfactorily in its electromagnetic energy (emissions) and be capable of offering adequate protection (immunity) against such energy occurring in the environment.

The EMC Directive (89/336/EEC) as amended by Directives 92/31/EEC and 93/68/EEC became effective on 1 January 1996 and controls the emissions and immunity characteristics of industrial products.

The main requirement of this Directive is to ensure that the use of electronic equipment will not interfere with (or be interfered by) the operation of any other equipment.

Note: This Directive defines only protection requirements relating to electromagnetic compatibility.

4.3.2.2 Structure

The Directive includes the Annexes listed in Table 4.3.

Annex	*Title*
I	EC Declaration of Conformity EC Conformity Mark
II	Criteria for the assessment of the bodies to be notified
III	Illustrative list of the principal protection requirements

Table 4.3 Annexes to Directive 39/336/EEC

4.3.2.3 Objectives

The objective of 89/336/EEC is

> to harmonise the electromagnetic compatibility (EMC) protection requirements (including emission and immunity) in order to guarantee the operation and compatibility of apparatus in their intended EMC environment and to guarantee the free movement of apparatus in the EEA territory.

The Directive applies to all electrical and electronic appliances, equipment and installations containing electrical and/or electronic components that are liable to cause electromagnetic disturbance or where the performance may be affected by an electromagnetic disturbance.

This Directive also applies to electrical household appliances, portable tools and similar equipment and fluorescent lighting luminaires fitted with starters.

The principal protection requirements are set out in Annex III of the EMC Directive.

4.3.2.4 Exclusions

The following are totally excluded from the requirements of the EMC Directive:

- active implantable medical devices;
- equipment for aircraft in flight;
- equipment such as cables, cabling accessories, connectors etc.;
- medical equipment;
- motor vehicles;
- radio-amateur equipment not available commercially.

Apparatus excluded for the **immunity part** of the EMC Directive include:

- non-automatic weighing instruments.

Apparatus excluded for the **emission part** of the EMC Directive include:

- agricultural and forestry tractors.

4.3.2.5 Essential Requirements

The Directive defines (in Article 4) the Essential EMC Requirements in two general statements related to the emission of disturbances and the immunity to disturbances as follows:

> emission limitation to permit other apparatus to operate as intended;
> immunity requirements to permit the concerned apparatus to operate as intended.

and states that

> 'apparatus shall be so constructed that:
> the electromagnetic disturbance it generates does not exceed a level allowing radio and telecommunications equipment and other apparatus to operate as intended;
> the apparatus has an adequate level of intrinsic immunity of electromagnetic disturbance to enable it to operate as intended.'

The maximum electromagnetic disturbance generated by the apparatus shall be such as not to hinder the use of (in particular) the following apparatus:

- aeronautical and marine radio apparatus;
- domestic appliances and household electronic equipment;
- domestic radio and television receivers;
- educational electronic equipment;
- industrial manufacturing equipment;
- information technology equipment;
- lights and fluorescent lamps;
- medical and scientific apparatus;
- mobile radio equipment;
- mobile radio and commercial radiotelephone equipment;
- radio and television broadcast transmitters;
- telecommunications networks and apparatus.

4.3.2.6 Proof of Conformity

Apparatus conforming to this Directive shall be certified by an EC Declaration of Conformity that is issued by the manufacturer. The declaration shall be held at the disposal of the Competent Authority for 10 years following the placing of the apparatus on the market.

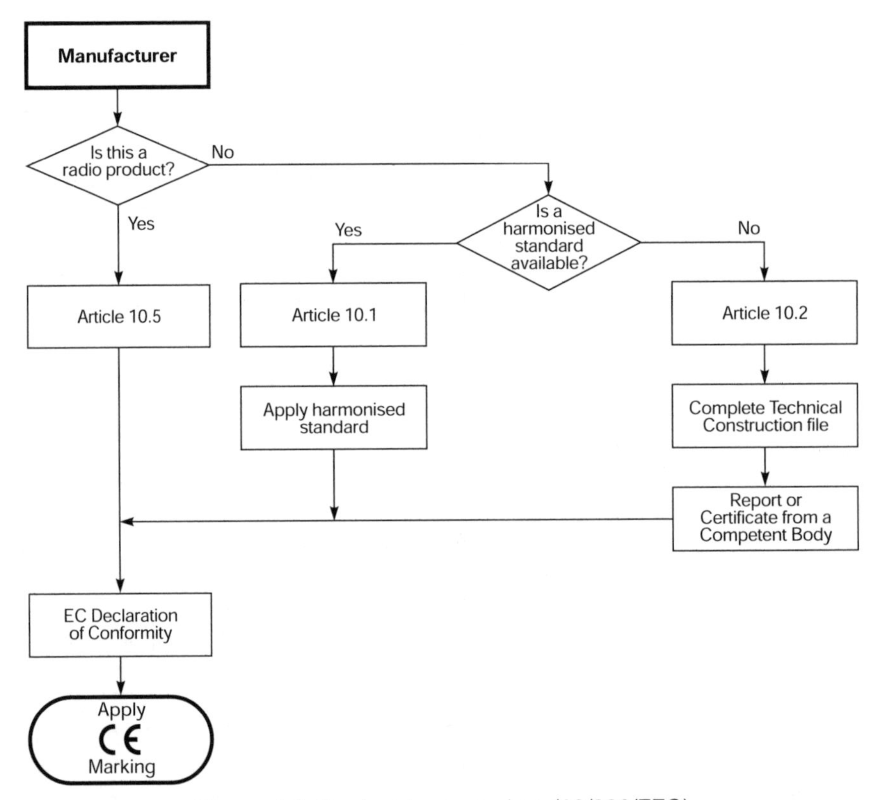

Figure 4.3 Conformity procedure (89/336/EEC)

Title	*Yes/No*
CE Marking	Yes
EC Declaration of Conformity	Yes
EC-Type Examination	Yes (Abstract 10.3)
Technical File	Yes (Abstract (10.2)

Table 4.4 Proof of Conformity (89/336/EEC)

The manufacturer shall also affix the EC conformity mark to the apparatus. Alternatively it can be affixed to the packaging, instructions for use, or the guarantee certificate.

Article 10 of the Directive specifies three procedures for the assessment of the conformity of apparatus. These are:

1 Apparatus for which the manufacturer has applied the harmonised standards (Article 10.1). The manufacturer draws up a written EC Declaration of Conformity stating the applied standards and affixes the CE Marking.
2 Where the manufacturer has not applied the standards or only applied them in part or there are no relevant standards (Article 10.2), a Technical Construction File must be established by the manufacturer and kept at the disposal of the Competent Authorities. The Technical File must include a technical report or certificate obtained from an independent or competent body in the Directive.
3 A specific procedure for apparatus that has been designed for the transmission of radio communications (Article10.5). Before issuing the Declaration of Conformity and affixing the CE Marking, the manufacturer must obtain an EC-Type Examination certificate issued by one of the Notified Bodies.

4.3.2.7 Other Relevant Information

From 1 January 1996, it has been agreed (by the Council of European Communities) that all active electronic devices **shall** comply with the EMC Directive and **only** equipment carrying the CE Marking may be offered for sale. By virtue of CE Marking, therefore, the user of such equipment has the supplier's guarantee of compliance with the EMC Directive.

4.3.2.7.1 Publications

- 'Guidelines on the application of the Council Directive 89/336/EEC', published by the Commission of the European Communities, Directorate General III (October 1993).
- 'Guidance on how to use the standards for the implementation of the EMC Directive' Reference: CLC(PERM) 009. Published as a Standing Document by CENELEC.
- 'Guide on EMC Standardisation for Product Committees' (Reference R1110–001), published by CENELEC.
- 'Guide to Generic Standards' (Reference R110–002), published by CENELEC.
- Application Guide (available from the Commission) 'Guidelines on the application of 89/336/EEC'.

Further details of the requirements made in this Directive and other associated harmonised Directives/standards are contained in another of my books entitled *Quality and Standards in Electronics*.

Detailed information concerning Environmental Standards is contained in *Environmental Requirements for Electromechanical and Electronic Equipment*.

Copies of these books are available from StingRay, Riddiford House, Winkleigh, Devon EX19 8DW, Fax (+44) 01837 83011 or via the website at **www.herne.demon.co.uk**.

4.3.2.7.2 CE Marking

Where apparatus is the subject of other Directives covering other aspects – and which also provide for the CE Marking – the latter shall indicate that the appliances are also presumed to conform to those other Directives. However, where one or more of these Directives allow the manufacturer, during a transitional period, to choose which arrangements to apply, the CE Marking shall indicate conformity only to the Directives applied by the manufacturer. In this case, particulars of the Directives applied, as published in the *Official Journal of the European Communities*, must be given in the documents, notices or instructions required by the Directives and accompanying such apparatus.

4.3.3 Safety of Machinery (98/37/EC)

Directive:	**98/37/EC** **Dated: 12 Jun 98**
Short name	Machinery
Description	Safety of all machines with moving parts
Amendments	98/79/EC
Repealed Directives	89/392/EEC 91/368/EEC (only Article 1) 93/44/EEC 93/68/EEC (only Article 6)
Member States Implementation Date	89/392/EEC, 1 Jan 92 91/368/EEC, 1 Jan 92 93/44/EEC, 1 Jul 94 93/68/EEC, 1 Jul 94 Implemented in the UK by the Supply of Machinery (Safety) Regulations 1992 (SI 1992/3072)
Transitional Period	Not applicable
Date of entry into force	For those previously covered by: 89/392/EEC, 1 Jul 95 91/368/EEC, 1 Jan 93 93/44/EEC, 1 Jan 95 93/68/EEC, 1 Jan 95

4.3.3.1 Introduction

It is a well-known fact that the machinery sector is a very important part of the engineering industry and one of the industrial mainstays of the EU economy. Unfortunately there are an increasing number of accidents caused through the use of machinery; but through the introduction of safe design, safe construction, proper installation and adequate maintenance, an attempt is being made to improve this situation.

This Directive applies to all machinery and lays down (in Annex 1) the essential health and safety requirements, supplemented by a number of more specific requirements for certain categories of machinery. This Directive also applies to safety components (i.e. a component fulfilling a safety function when in use and the failure or malfunctioning of which endangers the safety or health of exposed persons) placed on the market separately.

Over the years, the previous Machinery Directive (89/392/EEC dated of 14 June 1989) had been frequently and substantially amended and so for reasons of clarity and rationality it was decided to completely reformat the said Directive.

4.3.3.2 Structure

The Directive includes the Annexes listed in Table 4.5.

4.3.3.3 Objectives

To harmonise the national regulations concerning the design, manufacture and supply of machinery by defining Essential Requirements to ensure its safe use.

4.3.3.4 Exclusions

In the context of the Machinery Directive (MSD), a machine is 'an assembly of linked parts, at least one of which moves'. It is 'the assembly of linked parts or components, at least one of which moves, with the appropriate actuators, control and power circuits etc., joined together for a specific application, in particular for the processing, treatment, moving and packaging of a material'.

Annex	*Title*
I	Essential Health and Safety Requirements relating to the design and construction of machinery and safety components
II	Contents of the EC Declaration of Conformity
III	CE Conformity Marking
IV	Types of machinery and safety components for which the procedure referred to in Article 8(2)(b) and (c) must be applied
V	EC Declaration of Conformity
VI	EC-Type Examination
VII	Minimum criteria to be taken into account by Member States for the Notification of Bodies
VIII	Repealed Directives Lists of deadlines for transposition into and application in national law
IX	Correlation Table (between 89/332/EEC and This Directive)

Table 4.5 Annexes to Directive 98/37/EC

The term 'machinery' also:

- covers

  'an assembly of machines which, in order to achieve the same end, are arranged and controlled so that they functional an integral whole';

- means

  'interchangeable equipment modifying the function of the machine, which is placed on the market for the purpose of being assembled with a machine or a series of different machines or with a tractor by the operator himself in so far as this equipment is not a spare part or a tool'.

Such a wide definition leaves room for little exclusion! However, there is a list of specific categories of machines that are specifically excluded such as:

- agricultural or forestry tractors (defined in Article 1(1) of 74/150/EEC of 4 Mar 74 as amended by 88/297/EEC);
- cableways, including funicular railways for the public or private transportation of persons;
- construction site hoists intended for the lifting of persons or persons and goods;
- lifts which serve specific buildings and constructions having a car moving between guides which are rigid and inclined at an angle of more than 15 degrees to the horizontal and designed for the transport of persons, persons and goods, or goods alone;
- machinery whose only power source is directly applied through manual effort unless it is a machine for lifting or lowering loads;
- machinery for medical use used in direct contact with patients;
- machinery specially designed or put into service for nuclear purposes which, in the event of failure, may result in the emission of radioactivity;
- rack and pinion mounted vehicles for transporting people;
- means of transport of vehicles and their trailer intended solely for transporting passengers by air or on the road, rail or water networks;
- mine winding gear;
- radioactive sources forming part of a machine;
- firearms;
- seagoing vessels and mobile offshore units together with equipment on board such vessels or units;
- steam boilers, tanks and pressure vessels;
- storage tanks and pipelines for petrol, diesel fuels inflammable liquids and dangerous substances;
- special equipment for use in fairgrounds and/or amusement parks;

- those machines that are mainly of electrical origin shall be covered by the Low Voltage Directive (i.e. 73/23/EEC);
- theatre elevators.

Note: Other Directives are available (or will shortly be available) to cover these exclusions in general terms.

Although machines supplied before 1 January 1993 are exempt from the new regulations, if the machinery is substantially re-engineered (in such a way that the risk profile is significantly altered) and then the machine is re-supplied second hand, the modified machines should meet the new regulations, and CE Marking **is** required.

4.3.3.5 Essential Requirements

These are contained in Annex 1 which is split up into a number of sections as follows:

4.3.3.5.1 Essential Health and Safety Requirements

Covering:

- principles of safety integration;
- controls;
- protection against mechanical hazards;
- required characteristics of guards and protection devices;
- protection against other hazards;
- electricity supply;
- static electricity;
- extreme temperatures;
- fire;
- explosion;
- noise;
- vibration;
- radiation;
- emission of dust, gases, etc.;
- maintenance;
- indicators.

Note: Full details of Environmental Requirements for Electromechanical and Electronic equipment are contained in my recent book of the same title (see reference section for further details).

4.3.3.5.2 Essential Health and Safety Requirements for Certain Categories of Machinery

Covering:

- agri-foodstuffs machinery;
- portable hand-held and/or hand-guided machinery;
- machinery for working wood and analogous materials.

4.3.3.5.3 Essential Health and Safety Requirements to Offset the Particular Hazards due to the Mobility of Machinery

Including:

- work stations;
- controls;
- protection against mechanical hazards;
- protection against other hazards.

4.3.3.5.4 Essential Health and Safety Requirements to Offset the Particular Hazards due to a Lifting Operation

- protection against mechanical hazards;
- special requirements for machinery whose power source is other than manual effort;
- CE Marking;
- instruction handbook.

4.3.3.5.5 Essential Health and Safety Requirements for Machinery Intended for Underground Work

- risks due to lack of stability;
- movement;
- lighting;
- control devices;
- stopping;
- fire;
- emissions of dust, gases, etc.

4.3.3.5.6 Essential Health and Safety Requirements to Offset the Particular Hazards due to the Lifting or Moving of Persons

- controls;
- risks of persons falling from the carrier;
- risks of the carrier falling or overturning;
- CE Marking.

In Annex 1, the Machinery Directive also lists the essential health and safety regulations that machinery is required to meet before CE Marking can be affixed. These address all hazards presented by machinery to operators.

Similar to other New Approach Directives, the essential health and safety requirements laid down in this Directive are mandatory.

4.3.3.6 Proof of Conformity

Member States shall regard the following as conforming to all the provisions of this Directive, including the procedures for checking the conformity provided for in Chapter II:

- machinery bearing the CE Marking and accompanied by the EC Declaration of Conformity;
- safety components accompanied by the EC Declaration of Conformity.

In order to certify the conformity of machinery with the Essential Requirements in the annexes to the Directive, the manufacturer has to draw up a technical construction file and:

- for normal machinery, the manufacturer must affix to the machine the CE Marking;
- for machinery referred to in Annex IV (e.g. circular saw, presses etc.), the manufacturer must draw up an EC Declaration of Conformance.

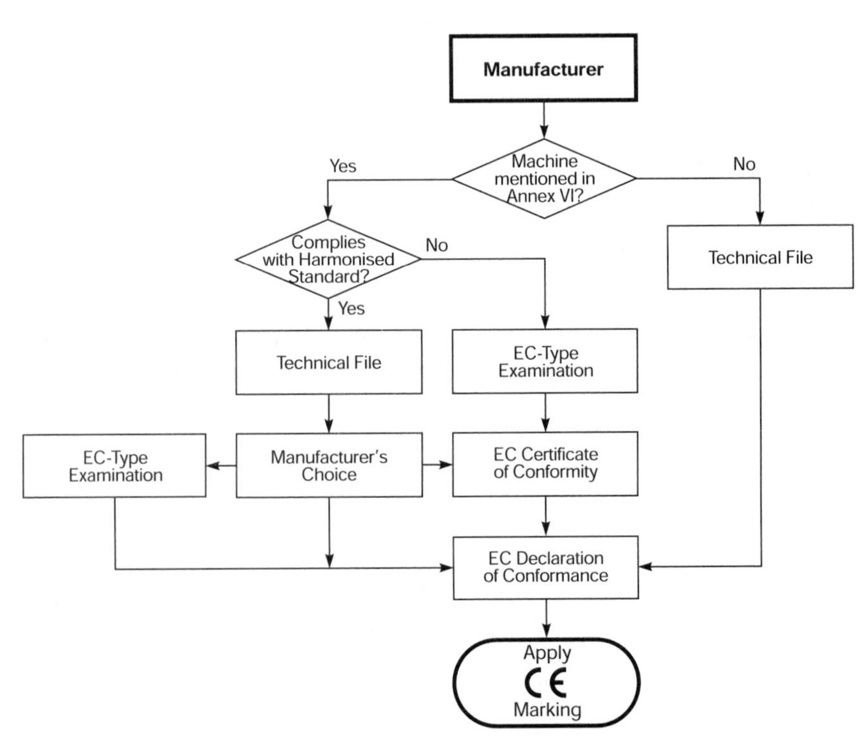

Figure 4.4 Conformity Procedure (98/37/EEC)

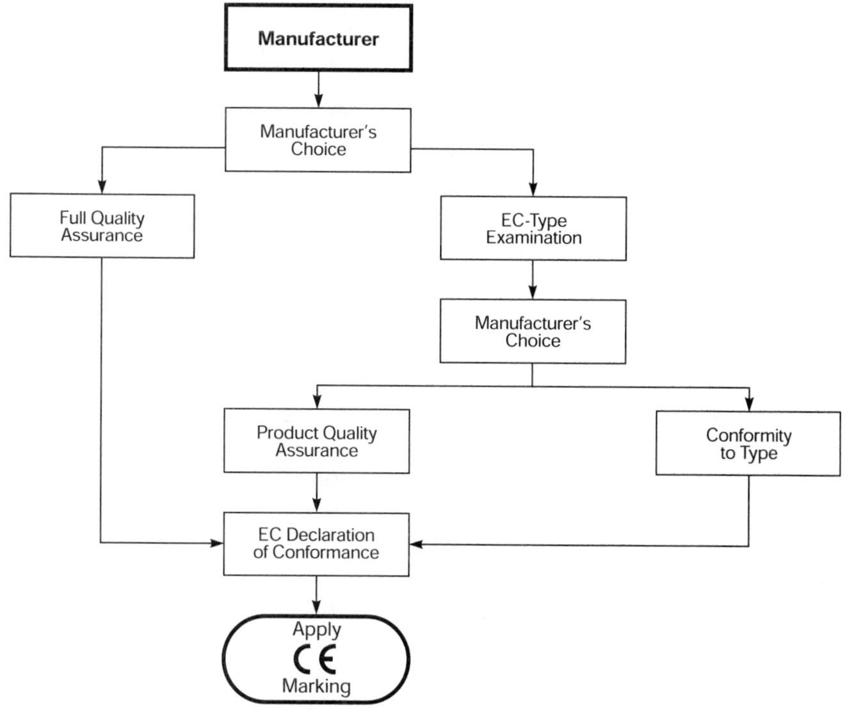

Figure 4.5 Safety Conformance procedure (98/37/EEC)

Title	*Yes/No*
CE Marking	Yes
EC Declaration of Conformity	Yes
EC-Type Examination	Yes
Product Quality Assurance	Yes*
Full Quality Assurance	Yes*
Technical File	Yes
Internal Production Control	
Conformity to Type	Yes*

* Manufacturer's choice

Table 4.6 Proof of Conformity (98/37/EC)

- If the machinery is referred to in Annex IV and is manufactured in accordance with a recognised standard(s), the manufacturer either:
  - draws up EC-Type Examination;
  - submits the Technical File to the Notified Body for verification and approval; or
  - submits an example of the machinery for an EC-Type Examination.

If machinery has a higher risk factor, then a guarantee of quality is also required in support of the EC-Type Examination procedure.

4.3.3.7 Other Relevant Information

It is necessary not only for Member States to ensure the free movement of machinery bearing the CE Marking and having an EC conformity certificate but also to ensure the free movement of machinery **not** bearing the CE Marking **when** it has been incorporated into other machinery or assembled with other machinery to form a complex installation.

This Directive defines only the essential health and safety requirements of general application, supplemented by a number of more specific requirements for certain categories of machinery.

4.3.3.7.1 Publications

- *Community legislation on machinery – Comments on Directive 89/392/EEC and Directive 91/368/EEC.* ISBN 92-826-5692-6 P. Massimi and J.-P. Van Gheluwe, English. EUR-OP. Luxembourg. 1993. (**Note**: This book was originally published also in Danish, Dutch, French, German, Greek, Italian, Portuguese and Spanish under different ISBN numbers.)
- *Standardisation programme – Safety of machinery,* 3rd edition, 1996/01, ISBN 92-9097-528-8. English. CEN, Brussels, 1996.

4.3.3.8 CE Marking

The Directive specifically states that:

- all machinery must be marked legibly and indelibly with the following minimum particulars:
  - name and address of the manufacturer;
  - CE Marking consisting of the initials ‘CE’;
  - designation of series or type;
  - serial number if any.
- if the machine is intended for use in a potentially explosive atmosphere, this must be indicated on the machinery;
- machinery must also bear full information relevant to its type and essential to its safe use (e.g. maximum speed of certain rotating parts, maximum diameter of tools to be fitted, mass etc.);

- where a machine part must be handled during use with lifting equipment, its mass must be indicated legibly, indelibly and unambiguously.

Notes: It is now no longer required to put the date next to the CE Marking although the year of construction must be shown elsewhere on the machine.

The minimum vertical dimension of 5 mm may be waived for small-scale machinery.

4.3.4 Equipment and Protective Systems in Potentially Explosive Atmospheres (94/9/EC)

Directive:	**94/9/EC** **Dated: 23 Mar 1994**
Short name	ATEX 100A
Description	Safety requirements for control systems and equipment for use in explosive atmospheres (e.g. coal mines)
Amendments	Nil
Repealed Directives	76/117/EEC (Surface Industry) 79/196/EEC (as amended by 90/487/EEC) 82/130/EEC (Gassy mines)
Member States Implementation Date	1 Mar 96 No additional UK legislation required for implementation
Transitional Period	The existing Directives (i.e. 76/117/EEC and 82/130/EEC) will run in parallel with ATEX 100A until withdrawn on 1 Jul 03
Date of entry into force	30 Jun 03

4.3.4.1 Introduction

Previously Framework Directive 76/117/EEC (concerning electrical equipment for use in potentially explosive atmospheres) and 82/130/EEC (concerning electrical equipment for use in potentially explosive atmospheres in mines susceptible to firedamp which only related to electrical equipment) were available. Although the protective measures and the test methods are often very similar, if not identical, for both mining and surface equipment it was agreed that a single Directive covering all protective equipment systems intended for use in potentially explosive atmospheres was required. Hence, 94/9/EC.

This Directive, therefore, covers both mining and non-mining equipment and is different to the previous Directives in that it includes mechanical as well as electrical equipment. It also includes explosive risks due to dust in the air as well as gases, vapours and mists in air which were excluded in the previous Directive. The new Directive, which is commonly called the ATEX 100A Directive, came into effect on a voluntary basis on 1 March 1996 and will be mandatory from 1 July 2003.

4.3.4.2 Structure

The Directive includes the Annexes listed in Table 4.7.

4.3.4.3 Objectives

To remove barriers to trade between the Member States concerning equipment and protective systems intended for use in potentially explosive atmospheres.

Annex	*Title*
I	Criteria determining the classification of Equipment-Groups into categories
II	Essential health and safety requirements relating to the design and construction of equipment and protective systems intended for use in potentially explosive atmospheres
III	EC-Type Examination
IV	Production Quality Assurance
V	Product Verification
VI	Conformity to Type
VII	Product Quality Assurance
VIII	Internal Control of Production
IX	Unit Verification
X	CE Marking Content of the EC Declaration of Conformity
XI	Minimum criteria to be taken into account by Member States for the Notification of Bodies

Table 4.7 Annexes to Directive 94/9/EC

4.3.4.3.1 Groups

Two groups of equipment are defined in this Directive:

- **Group I** applies to equipment intended for use in underground parts of mines and to those parts of surface installations of such mines that are liable to be endangered by firedamp and/or combustible dust.
- **Group II** applies to equipment intended for use in other places liable to be endangered by explosive atmospheres.

4.3.4.3.2 Categories

The Directive classifies equipment into five categories depending on the equipment's area of use. In descending level of use, these are:

- **Category M1** equipment specifically intended for mining use and required to remain functional in the presence of an explosive atmosphere.
- **Category M2** equipment is also intended for mining use but is intended to be de-energised in the event of an explosive atmosphere.
- **Category 1**, non-mining equipment intended for use where an explosive atmosphere is present continuously, for long periods or frequently.
- **Category 2**, non-mining equipment intended for use in areas where an explosive atmosphere is likely to occur.
- **Category 3**, non-mining equipment intended for use in areas where an explosive atmosphere is unlikely to occur or is likely to do so only infrequently and for short periods.

4.3.4.4 Exclusions

The following are excluded from the scope of this Directive:

- medical devices intended for use in a medical environment;
- equipment and protective systems where the explosion hazard results exclusively from the presence of explosive substances or unstable chemical substances;
- equipment intended for use in domestic and non-commercial environments where potentially explosive atmospheres are rarely created (i.e. solely as a result of the accidental leakage of fuel gas);
- personal protective equipment covered by Directive 89/686/EEC;
- seagoing vessels and mobile offshore units together with equipment on board such vessels or units;
- vehicles and their trailers intended solely for transporting passengers by air or by road, rail or water networks;
- vehicles intended for use in a potentially explosive atmosphere shall **not** be excluded;
- equipment covered by Article 223 (1) (b) of the Treaty.

4.3.4.5 Essential Requirements

Category M1, M2, 1 and 2 equipment must be capable of functioning, in conformity with the operational parameters established by the manufacturer, to a very high level of protection. For Category 3 the manufacturer must ensure a normal level of protection.

The Essential Requirements are fully described over eight pages in Annex II of the Directive and cover, in considerable detail, three basic areas.

4.3.4.5.1 Common Requirements for Equipment and Protective Systems

Consisting:

- general requirements;
- selection of materials;
- design and construction;
- potential ignition sources;
- hazards arising from external effects;
- requirements in respect of safety related devices;
- integration of safety requirements relating to the system.

4.3.4.5.2 Supplementary Requirements in Respect of Equipment

This area is further divided into four sections covering:

- Group I, Category M equipment;
- Group II, Category 1 equipment;
- Group II, Category 2 equipment;
- Group II, Category 3 equipment.

4.3.4.5.3 Supplementary Requirements in Respect of Protective Systems

This covers;

- general requirements;
- planning and design.

4.3.4.6 Proof of Conformity

As shown in Table 4.8, the procedures for assessing the conformity of equipment vary according to the Equipment Group.

4.3.4.6.1 Equipment Group I and II, Equipment Category M1 and 1

The manufacturer shall affix the CE Marking and follow the EC-Type Examination procedure in conjunction with:

- the procedure relating to Production Quality Assurance (Annex IV); or
- the procedure relating to Product Verification (Annex V).

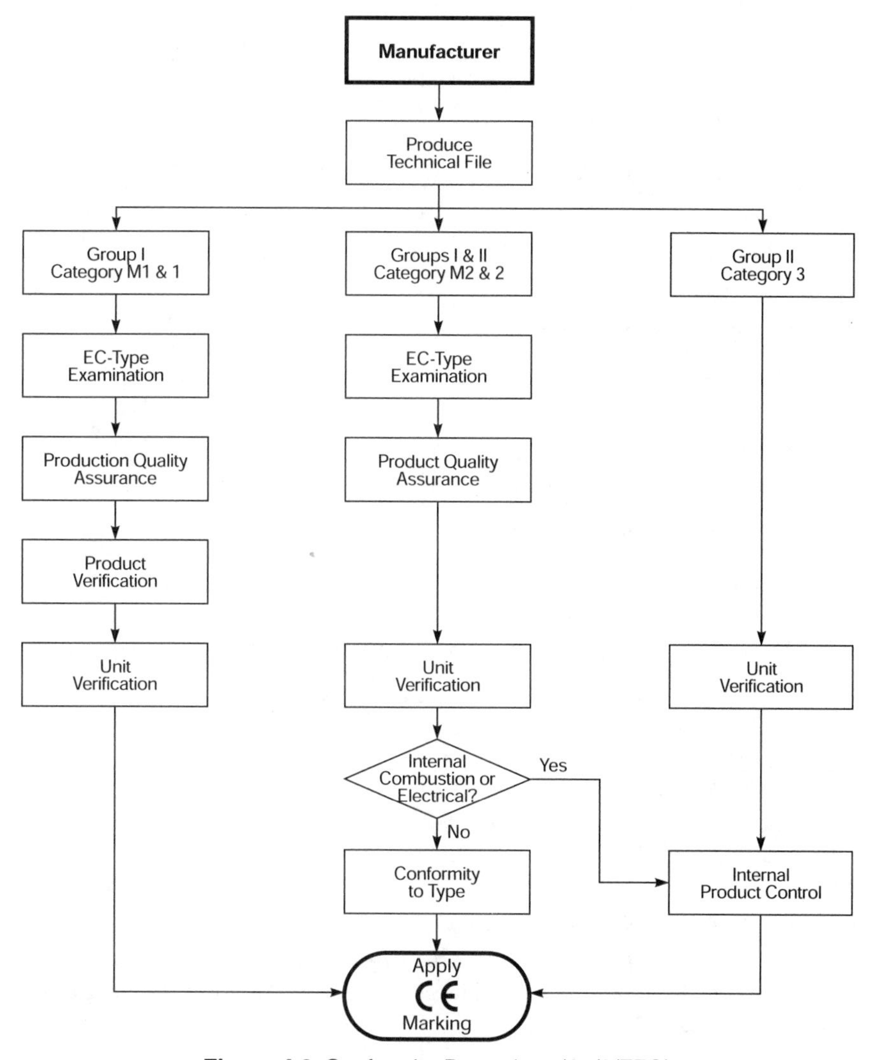

Figure 4.6 Conformity Procedure (94/9/EEC)

4.3.4.6.2 Equipment Group I and II, Equipment Category M2 and 2

For internal combustion engines and electrical equipment the manufacturer shall affix the CE Marking, follow the EC-Type Examination procedure, in conjunction with:

- the procedure relating to Conformity to Type; or
- the procedure relating to Product Quality Assurance.

For all other equipment, the manufacturer must affix the CE Marking and follow the procedure relating to internal control of production.

Title	*Group I & II category M1 & 1*	*Group I & II category M2 & 2**	*Group II category 3*
CE Conformity Marking	Yes	Yes	Yes
EC Declaration of Conformity	Yes	Yes	Yes
EC-Type Examination	Yes	Yes	
Product Quality Assurance		Yes	
Production Quality Assurance	Yes		
Product Verification	Yes		
Technical File	Yes	Yes	Yes
Unit Verification	Yes	Yes	Yes
Internal Production Control		*	Yes
Conformity to Type		Yes	

*Equipment other than internal combustion engines and electrical equipment need only follow the Internal Production Control route.

Table 4.8 Proof of Conformity (94/9/EEC)

4.3.4.6.3 Equipment Group II, Equipment Category 3

The manufacturer must affix the CE Marking and follow the procedure relating to internal control of production (Annex VIII).

4.3.4.6.4 Equipment Groups I and II

In addition to the procedures referred to above, the manufacturer may also, in order to affix the CE Marking, follow the procedure relating to CE Unit Verification (Annex IX).

4.3.4.7 Other Relevant Information

The intended use for equipment, protective systems and devices is based on information supplied by the manufacturer for their safe functioning in accordance with the equipment groups and categories.

4.3.4.8 CE Marking

The Directive specifically states that equipment and protective systems must be marked legibly and indelibly with the following minimum particulars:

- name and address of the manufacturer;
- CE Marking;
- designation of series or type;
- serial number, if any;
- year of construction;
- the specific Marking of explosion protection (see Figure 4.7) followed by the symbol of the equipment group and category;

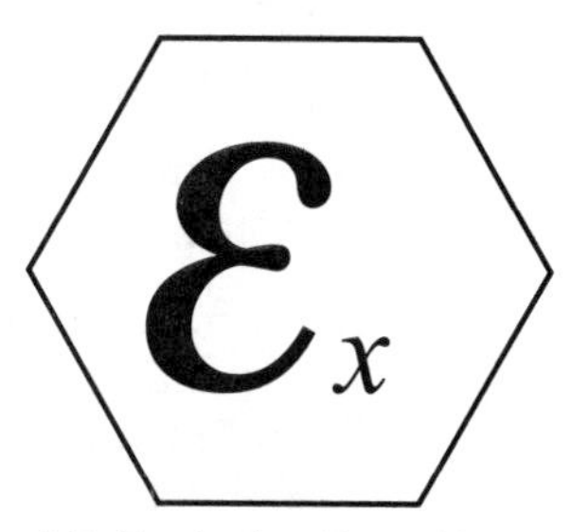

Figure 4.7 Explosion Protection Marking

- for equipment-group II, the letter 'G' (concerning explosive atmospheres caused by gases, vapours or mists); and/or
- the letter 'D' (concerning explosive atmospheres caused by dust).

Furthermore, where necessary, they must also be marked with all the relevant information essential for their safe use.

4.3.5 Radio and Telecommunications Terminal Equipment (99/5/EC)

Directive:	**99/5/EC Dated: 9 March 1999**
Short name	RTTE
Amendments	Nil
Repealed Directives	91/263/EEC (Telecommunication Terminal Equipment) 93/97/EEC (Satellite Earth Station Equipment) 98/13/EEC (Telecommunication Terminal Equipment and Satellite Earth Station Equipment) . . . from 8 April 2000
Member States Implementation Date	7 April 2000
Transitional Period	8 April 2002 provided that the equipment was previously in accordance with 98/13/EC
Date of entry into force	1 Jan 97

4.3.5.1 Introduction

Prior to 1998, telecommunications industrial products generally fell within the scope of four Directives, namely:

- the Telecommunications Terminal Equipment (TTE) Directive;
- the Satellite Earth Station Equipment (93/97/EEC);
- the ElectroMagnetic Compatibility (EMC) Directive (89/336/EEC);
- the Low Voltage (LVD) Directive 72/23/EEC.

As the requirements for Telecommunications Terminal Equipment and Satellite Earth Station Equipment were very similar, it was decided to combine all of the requirements for Telecommunications Terminal Equipment and Satellite Earth Station Equipment into one Directive – 98/13/EEC.

Under 93/13/EC, the Commission was required to draw up every second year a report on the implementation of this Directive, including progress on drawing up the relevant harmonised standards and on transforming them into technical regulations, as well as any problems that have arisen in the course of implementation. The report confirmed that It was the decision of the Commission that as 93/13/EC was very cumbersome and in some circumstances over prescriptive it needed to be completely revised.

As the proposed revisions were so extensive, it was decided to make a new Directive (i.e. 99/5/EC) – **except** that protection and safety requirements and certain conformity assessment procedures of Directives 73/23/EEC and 89/336/EEC should still remain in force.

It was agreed that the Commission would review the operation of this Directive and report (to the Council) not later than 7 October 2000 and every third year thereafter.

4.3.5.2 Structure

The Directive includes the Annexes Listed in Table 4.9.

4.3.5.3 Objectives

Quote from European Commission DG13:

> The RTTE Directive constitutes a fundamental change to the regulations in the Community that govern Radio and Telecommunications Terminal equipment. Besides replacing and relaxing the existing pan-European approval regimes as laid down by Directive 98/13/EC, the Directive also replaces all national approval regimes.

Annex	*Title*
I	Equipment not covered by this directive as referred to in article 1(4)
II	Conformity assessment procedure Module A (Internal Production Control)
III	Conformity assessment procedure (Internal production control plus specific apparatus tests)
IV	Conformity assessment procedure (Technical construction file)
V	Conformity assessment procedure (Full quality assurance)
VI	Minimum criteria to be taken into account by Member States when designating Notified Bodies
VII	Marking of equipment

Table 4.9 Annexes to Directive 99/5/EC

It is widely accepted that the radio and telecommunications terminal equipment (RTTE) sector is an essential part of the telecommunications market and a key element of the economy in the EU. It has therefore been given priority status. Unfortunately, the current Directives for TTE were no longer capable of accommodating changes caused by new technology, market developments and network legislation. They were difficult to comply with and they tended to favour the Network Operators at the expense of Equipment Suppliers. There were also still many 'grey' products. It was decided, therefore, to produce a new, more user friendly Directive.

This Directive (i.e. 99/5/EC) is an interim 'Full Harmonisation Directive' (meaning that no additional national approval requirements can be placed on equipment covered by the Directive) and it has the advantage of:

- greatly reduced compliance requirements – few technical requirements;
- allows self-declaration (similar to the EMC Directive, etc.) for non-radio equipment;
- has a limited involvement of a Notified Body for Radio Equipment.

99/5/EC applies to telegraph terminal equipment that is intended to be connected directly or indirectly to the public telecommunications networks. It also applies to radio equipment. Eventually (i.e. post 2000) it is intended that the RTTE Directive will be replaced by a further, final, Directive.

4.3.5.4 Exclusions

The following are excluded from this Directive:

- equipment designed in conformity with 73/23/EEC;
- equipment designed in conformity with 89/336/EEC;
- marine equipment covered by 96/98/EC;
- cabling and wiring;
- receive-only radio equipment intended to be used solely for the reception of sound and TV broadcasting services;
- civil aviation products, appliances and components;
- air-traffic-management equipment and systems;
- apparatus exclusively used for activities concerning public security, defence, State security and the activities of the State in the area of criminal law;
- radio equipment used by radio amateurs – unless the equipment is available commercially (i.e. in kit form for assembly by radio amateurs).

Note: Commercial equipment modified by and for the use of radio amateurs is not regarded as commercially available equipment.

4.3.5.5 Essential Requirements

The Essential Requirements relevant to a class of radio and telecommunications terminal equipment depends on the nature and the needs of that class of equipment. These requirements must be applied with discernment in order not to inhibit technological innovation or the meeting of the needs of a free-market economy.

Care should be taken that radio equipment and telecommunications terminal equipment should not represent an avoidable hazard to health.

Radio equipment and telecommunications terminal equipment should be designed in such a way that disabled people may use it without or with only minimal adaptation. Features that have been introduced on the radio equipment and telecommunications terminal equipment in order to prevent the infringement of personal data and privacy of the user should be permitted to work uninterruptedly.

Radio telecommunications terminal equipment shall not impair any built-in functions required by emergency services.

Public Telecommunications Networks (PTNs) should be able to define the technical characteristics of their interfaces, subject to the competition rules of the Treaty. Each Member State shall notify to the Commission the types of interface offered.

It should be possible to identify and add specific Essential Requirements on user privacy, features for users with a disability, features for emergency services and/or features for avoidance of fraud.

Unacceptable degradation of service to persons other than the user of radio equipment and telecommunications terminal equipment should be prevented.

Effective use of the radio spectrum should be ensured so as to avoid harmful interference.

- **Safety** – the objectives contained in Directive 73/23/EEC (LVD), with respect to safety requirements, but with no lower voltage limit applying.
- **EMC** – the protection requirements contained in Directive 89/336/EEC (EMC), with respect to electromagnetic compatibility.
- **Efficient Use of Spectrum** – radio equipment shall be so constructed that it effectively uses the spectrum allocated to terrestrial/space radio communication and orbital resources so as to avoid harmful interference.

4.3.5.6 Proof of Conformity

At the choice of the manufacturer, compliance of the apparatus with the Essential Requirements may be demonstrated by using the procedures specified in Directive 73/23/EEC and Directive 89/336/EEC respectively,

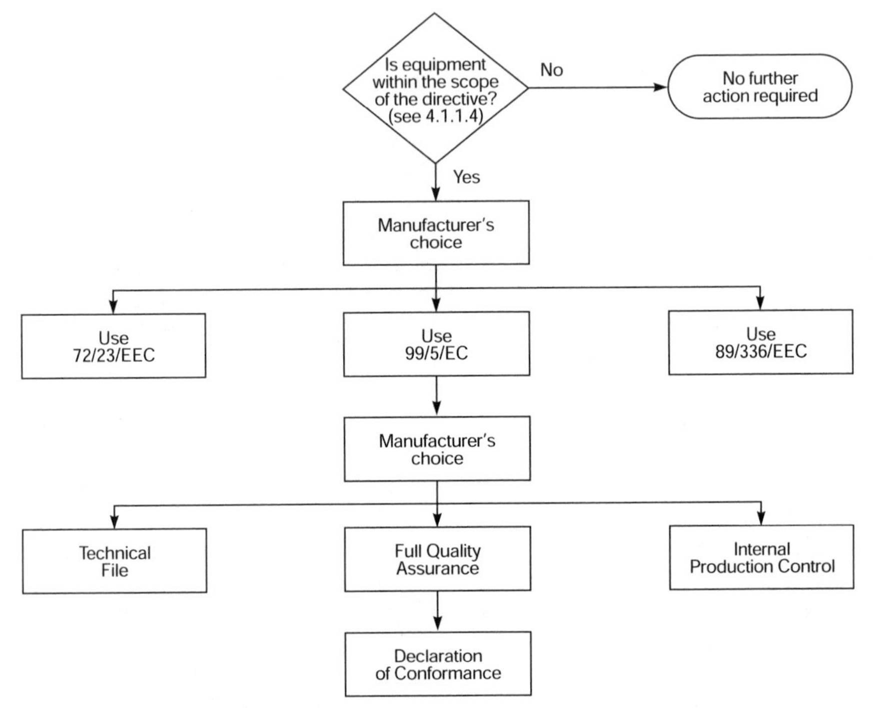

Figure 4.8 Conformity Procedure (99/5/EC)

Title	*Yes/No*
CE Marking	Yes
EC Declaration of Conformity	Yes
Full Quality Assurance	Yes[1]
Technical File	Yes[1]
Internal Production Control	Yes[1]

1 Manufacturer's choice for telecommunications terminal equipment which does not make use of the spectrum allocated to terrestrial/space radio communication and receiving parts of radio equipment.

Table 4.10 Proof of Conformity (99/5/EC)

where the apparatus is within the scope of those Directives as an alternative to the procedures laid out below.

For radio equipment other than telecommunications terminal equipment which does not make use of the spectrum allocated to terrestrial/space radio communication and receiving parts of radio equipment, the requirements for Internal Production Control also include the following additional tests:

- for each type of apparatus, essential radio test suites that have been identified by the Notified Body shall be carried out by the manufacturer;
- the manufacturer must declare that these tests have been carried out and that the apparatus complies with the Essential Requirements;
- radio equipment, in addition to the normal requirements for CE Marking, shall carry the equipment class identifier – where such identifier has been assigned.

4.3.5.7 Other Relevant Information

Under Directive 98/10/EC (regarding open network provision (ONP) for voice telephony), national regulatory authorities are required to publish details of their technical interface specifications for network access (this is to facilitate a competitive market for the supply of terminal equipment).

Dual-use goods are subject to the EU regime of export controls introduced by Council Regulation (EC) No 3381/94.

4.3.5.7.1 Publications

For further information, see:

- TAPC website: *http://www.tapc.org.uk/*
- Text of RTTE Directive: *http://www.tapv.org.uk/document/*
- RTTE Steering Group (official EC committee) *http://www.ispo.cec.be/infosoc/telecompolicy/type appr/steering.htm*
- DTI's Views and Comments: *http://www2.dti.gov.uk/eurobrief/3rtte.htm*
- Update from Radiocommunications Agency (June 1999) *http://www.open.gov.uk/radiocom/rtte/rtteweb.htm*

4.3.5.7.2 CE Marking

CE Marking is presumed to indicate compliance with all relevant directives. This means that all products carrying the CE Marking will be presumed to comply with the RTTE **if the directive is considered to apply to them**.

The assignment of radio equipment class identifiers (see 4.3.5.6) should draw on the expertise of CEPT/ERC and of the relevant European standards bodies in radio matters.

4.3.6 Personal Protective Equipment (89/686/EEC)

Directive:	**89/686/EEC Dated: 21 Dec 89**
Short name	Personal Protective Equipment (PPE)
Description	Performance of equipment designed to protect the user from injury
Amendments	93/68/EEC (CE Marking) 93/95/EEC (Lengthened transitional period and additional essential requirement)
Repealed Directives	89/686/EEC (Personal Protective Equipment)
Member States Implementation Date	1 Jul 92 Implemented in the UK by The Personal Protective Equipment (EC Directive) Regulations 1992 (SI 1992:3139)
Transitional Period	Ending 30 Jun 95 Member States have agreed to accept all PPE that complies with the regulations that were in force (in their territory) on 30 Jun 92
Date of entry into force	30 June 1995

4.3.6.1 Introduction

The Directive lays down the conditions governing the placing on the market, free movement within the EU, and the basic safety requirements which Personal Protective Equipment (i.e. any device or appliance designed to be worn or held by an individual for protection against one or more health or safety hazards) must satisfy. The Directive applies to PPE intended for professional and private use (e.g. sports, leisure, domestic use, etc.).

4.3.6.2 Structure

The Directive includes the Annexes Listed in Table 4.11.

Note: Annex II (i.e. basic safety requirements) defines:

- general requirements applicable to all PPE (ergonomics, innocuousness, comfort and efficiency, etc.);
- additional requirements specific to a particular PPE (e.g. equipment for eye protection must not restrict the field of vision);
- additional requirements specific to particular risks.

Annex	*Title*
I	PPE Classes not covered by this Directive
II	Basic Health and Safety requirements
III	Technical Documentation supplied by the Manufacturer
IV	CE Conformity Marking
V	Conditions to be fulfilled by the Bodies of which Notification has been given
VI	EC Declaration of Conformity

Table 4.11 Annexes to Directive 89/686/EEC

4.3.6.3 Objectives

To remove barriers to trade between Member States in personal protective equipment (PPE) by harmonising basic requirements for the design, manufacture, testing and certification of these goods whilst, at the same time, maintaining the highest possible level of protection and ensuring the health and safety of users. This Directive applies to all PPE intended for both professional and private use whether for sports, leisure or domestic use.

4.3.6.4 Exclusions

The following are excluded from this Directive:

- PPE designed and manufactured specifically for use by the armed forces or in the maintenance of law and order (helmets, shields, etc.);
- PPE for self-defence (aerosol canisters, personal deterrent weapons, etc.);

- PPE designed and manufactured for private use against:
  - adverse atmospheric conditions (headgear, seasonal clothing, footwear; umbrellas, etc.);
  - damp and water (dish-washing gloves, etc.);
  - heat (gloves, etc.).
- PPE intended for the protection or rescue of persons on vessels or aircraft;
- Helmets and visors intended for users of two wheeled motor vehicles – pending the introduction of specific requirements for such helmets.

4.3.6.5 Essential Requirements

Annex II of the Directive defines both the general requirements (e.g. ergonomics, innocuousness, comfort, efficiency, etc.) applicable to all PPE together with the more specific requirements that are applicable to a particular type of PPE (e.g. equipment for eye protection must not restrict the field of vision). It also covers:

- a unit made up of several devices or appliances for the protection of an individual;
- a protective device or appliance combined with personal non-protective equipment;
- interchangeable PPE components which are used exclusively for such equipment.

4.3.6.5.1 General Requirements Applicable to all PPE

PPE must provide adequate protection against all risks encountered. Specifically mentioned are the:

- design principles;
- innocuousness of PPE (i.e. absence of risks and other 'inherent' nuisance factors);
- comfort and efficiency;
- information supplied by the manufacturer.

4.3.6.5.2 Additional Requirements Common to Several Classes or Types of PPE

This covers PPE:

- incorporating adjustment systems;
- 'enclosing' the parts of the body to be protected;
- for the face, eyes and respiratory tracts;
- subject to ageing;
- which may be caught up during use;
- for use in explosive atmospheres;

- intended for emergency use or rapid installation and/or removal;
- for use in very dangerous situations;
- incorporating components which can be adjusted or removed by the user;
- for connection to another, external complementary device;
- incorporating a fluid circulation system;
- bearing one or more identification or recognition marks directly or indirectly relating to health and safety;
- in the form of clothing capable of signalling the user's presence visually;
- 'multi-risk' PPE.

4.3.6.5.3 Additional Requirements Specific to Particular Risks

Covering protection against:

- mechanical impact;
- (static) compression of one part of the body;
- physical injury (abrasion, perforation, cuts, bites);
- drowning (lifejackets, armbands, and lifesaving suits);
- harmful effects of noise;
- heat and/or fire;
- cold;
- electric shock;
- radiation protection;
- dangerous substances and infective agents;
- safety devices for diving equipment.

4.3.6.6 Proof of Conformity

Before placing a PPE model on the market, the manufacturer is required to put together a Technical File.

EC-Type Examination of product models prior to manufacture will be required for most PPE, although the simple declaration of the manufacturer is sufficient for most PPE providing protection against minimal risk. This category covers PPE intended to protect the wearer against:

- mechanical action whose effects are superficial (gardening gloves, thimbles, etc.);
- cleaning materials of weak action and easily reversible effects (gloves affording protection against diluted detergent solutions, etc.);
- risks encountered in the handling of hot components which do not expose the user to a temperature exceeding 50°C or to dangerous impacts (gloves, aprons for professional use, etc.);
- atmospheric agents of neither an exceptional nor an extreme nature (headgear, seasonal clothing, footwear, etc.);

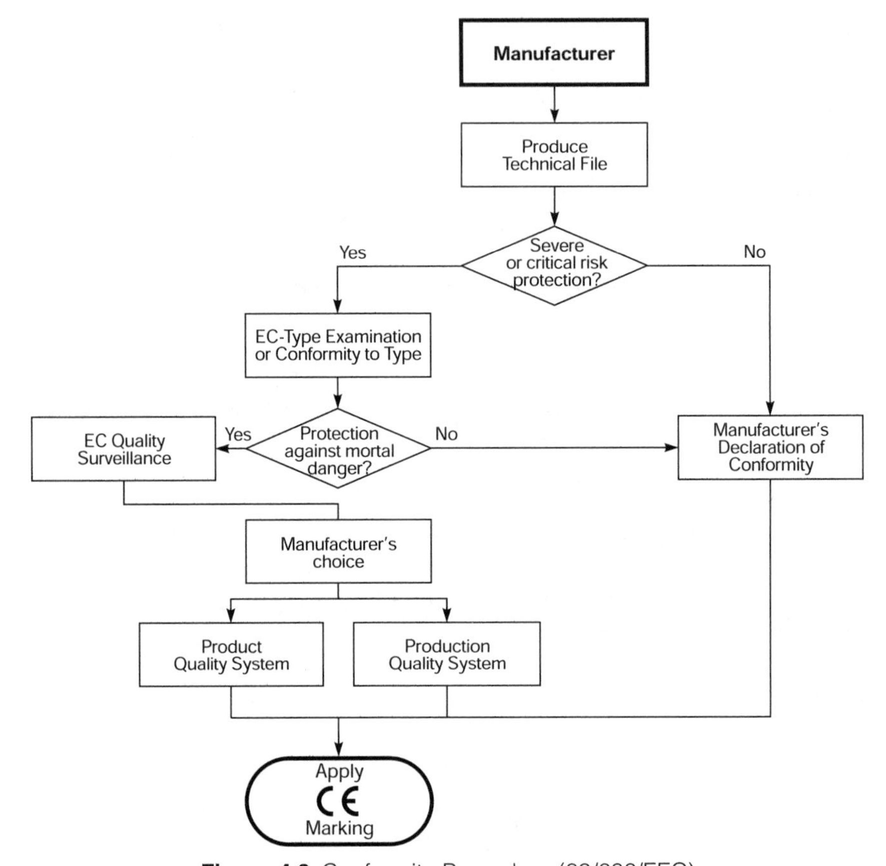

Figure 4.9 Conformity Procedure (89/686/EEC)

Title	*Yes/No*
CE Marking	Yes
EC Declaration of Conformity	Yes
EC-Type Examination	Yes
Product Quality Assurance	Yes
Production Quality Assurance	Yes
Technical File	Yes
EC Production Quality System	Yes
EC Final Product Quality System	Yes

Table 4.12 Proof of Conformity (89/686/EC)

- minor impacts and vibrations which do not affect vital areas of the body and whose effects cannot cause irreversible lesions (light anti-scalping helmets, gloves, light footwear, etc.);
- sunlight (sunglasses).

In the case of PPE providing protection against severe or even lethal risks, the basic procedure has to be supplemented by a surveillance of the production or the final product.

This category shall particularly cover:

- filtering respiratory devices for protection against solid and liquid aerosols or irritant, dangerous, toxic or radiotoxic gases;
- respiratory protection devices providing full insulation from the atmosphere, including those for use in diving;
- PPE providing only limited protection against chemical attack or against ionising radiation;
- emergency equipment for use in high-temperature environments the effects of which are comparable to those of an air temperature of 100°C or more and which may or may not be characterised by the presence of infra-red radiation, flames or the projection of large amounts of molten material;
- emergency equipment for use in low-temperature environments the effects of which are comparable to those of an air temperature of –50°C or less;
- PPE to protect against falls from a height;
- PPE against electrical risks and dangerous voltages or that used as insulation in high-tension work.

4.3.6.6.1 CE Marking

The following additional information regarding the affixing of CE Marking was made by the CE Amending Directive (93/68/EEC):

- Where an article is of a simple design (e.g. gardening gloves etc.), and where the user is capable of assuming a level of protection provided by the article, the minimum vertical dimension of 5 mm may be waived for small-scale PPE.
- The year, together with an identifying number of the Notified Body carrying out the control procedures, should be incorporated into CE Marking.
- The requirement to show the number of the body carrying out the EC-Type Examination is now no longer required.

4.3.7 Explosives for Civil Use (93/15/EEC)

Directive:	**93/15/EEC** **Dated: 19 Feb 73**
Short name	Civil Uses
Description	Performance and safety of commercial explosives excluding ammunition and pyrotechnics
Amendments	Nil
Repealed Directives	Nil
Member States Implementation Date	30 Jun 94 Implemented in the UK by the Placing on the Market and Supervision of Transfers of Explosives Regulations 1993 (SI 1993:2714)
Transitional Period	A transitional period ending on 31 December 2002 has been agreed. During this time explosives may be placed on the market provided that they conform with the national legislation that was in force before 31 December 1994
Date of entry into force	1 Jan 95

4.3.7.1 Introduction

This Directive defines only the Essential Requirements that must be met by conformity tests for explosives.

As the rules concerning the transport of explosives are already adequately covered by international conventions and agreements and (at international level) there are United Nations recommendations for the transport of dangerous goods (including explosives), the scope of this Directive does not concern the transport rules and/or the rules governing controls on transfers and associated arrangements.

Note: Transfers of ammunition need to be governed by provisions similar to those applicable to arms (see Directive 91/477/EEC on the control of the acquisition and possession of weapons) and work is currently progressing in this area.

4.3.7.2 Structure

The structure of the Directive is as shown in Table 4.13.

Annex	*Title*
I	Essential Safety Requirements
II	The Conformity Modules
II–1	EC-Type Examination
II–2	Module C – Conformity to Type
II–3	Module D – Production Quality Assurance
II–4	Module E – Product Quality Assurance
II–5	Module F – Product Verification
II–6	Module G – Unit Verification
III	Minimum criteria to be taken into account by Member States for the Notification of Bodies

Table 4.13 Annexes to Directive 93/15/EEC

4.3.7.3 Objectives

To lay down the basic requirements for the free movement of explosives for civil uses without lowering the optimum levels of safety (prevention of accidents or limitation of their effects) and security (prevention of illegal trade and use).

4.3.7.4 Exclusions

The Directive applies to explosives, i.e. materials and articles considered being such in the United Nations' *Recommendations on the Transport of Dangerous Goods* (the 'Orange Book') and falling within Class I of these recommendations.

This Directive shall not apply to:

- explosives, including ammunition, intended for use, in accordance with national law, by the armed forces or the police;
- pyrotechnical articles;
- ammunition with certain exceptions (e.g. as provided in Articles 10, 11, 12, 13, 17, 18 and 19 of the Directive).

4.3.7.5 Essential Requirements

Explosives falling within the scope of this Directive must comply with the essential safety requirements set out in Annex I which apply to them.

4.3.7.5.1 General Requirements

Each explosive must be designed, manufactured and supplied in such a way as to present a minimal risk to the safety of human life and health, and to prevent damage to property and the environment under normal, foreseeable conditions, in particular as regards the safety rules and standard practices including until such time as it is used.

Each explosive must attain the performance characteristics specified by the manufacturer in order to ensure maximum safety and reliability.

Each explosive must be designed and manufactured in such a way that when appropriate techniques are employed it can be disposed of in a manner which minimises effects on the environment.

4.3.7.5.2 Special Requirements

As a minimum, the following information and properties – where appropriate – must be considered:

- each explosive should be tested under realistic conditions (if this is not possible in a laboratory, the tests should be carried out in the conditions in which the explosive is to be used);
- construction and characteristic properties, including chemical composition, degree of blending and, where appropriate, dimensions and grain size distribution;
- the physical and chemical stability of the explosive in all environmental conditions to which it may be exposed;
- sensitiveness to impact and friction;
- compatibility of all components as regards their physical and chemical stability;
- the chemical purity of the explosive;
- resistance of the explosive against influence of water where it is intended to be used in humid or wet conditions and where its safety or reliability may be adversely affected by water;
- resistance to low and high temperatures, where the explosive is intended to be kept or used at such temperatures and its safety or reliability may be adversely affected by cooling or heating of a component or of the explosive as a whole;
- the suitability of the explosive for use in hazardous environments (e.g. environment endangered by firedamp, hot masses, etc.) if it is intended to be used under such conditions;
- safety features intended to prevent untimely or inadvertent initiation or ignition;

- the correct loading and functioning of the explosive when used for its intended purpose;
- suitable instructions and, where necessary, markings in respect of safe handling, storage, use and disposal in the official language or languages of the recipient Member State;
- the ability of the explosive, its covering or other components to withstand deterioration during storage until the 'use by' date specified by the manufacturer;
- specification of all devices and accessories needed for reliable and safe functioning of the explosive.

The various groups of explosives must at least also comply with the following requirements:

4.3.7.5.3 Blasting Explosives

The proposed method of initiation must ensure safe, reliable and complete detonation (or deflagration as appropriate of the blasting explosive). In the particular case of black powder, it is the capacity as regards deflagration that shall be checked.

Blasting explosives in cartridge form must transmit the detonation safely and reliably from one end of the train of cartridges to the other.

The gases produced by blasting explosives intended for underground use may contain carbon monoxide, nitrous gases, other gases, vapours or airborne solid residues only in quantities which do not impair health under normal operating conditions.

4.3.7.5.4 Detonating Cords, Safety Fuses, Igniter Cords and Shock Tubes

The covering of detonating cords, safety fuses and igniter cords must be of adequate mechanical strength and adequately protect the explosive filling when exposed to normal mechanical stress.

The parameters for the burning times of safety fuses must be indicated and must be reliably met.

Detonating cords must be capable of being reliably initiated, be of sufficient initiation capability and comply with requirements as regards storage even in particular climatic conditions.

4.3.7.5.5 Detonators (Including Delay Detonators) and Relays

Detonators must reliably initiate the detonation of the blasting explosives which are intended to be used with them under all foreseeable conditions of use.

Relays must be capable of being reliably initiated. The initiation capability must not be adversely affected by humidity.

The delay times of delay detonators must be sufficiently uniform to ensure that the probability of overlapping of the delay times of adjacent time steps is insignificant.

The electrical characteristics of electric detonators must be indicated on the packaging (e.g. no-fire current, resistance, etc.).

The wires of electric detonators must be of sufficient insulation and mechanical strength including the solidity of the link to the detonator, taking account of their intended use.

4.3.7.5.6 Propellants and Rocket Propellants

These materials must not detonate when used for their intended purpose.

Propellants where necessary (e.g. those based on nitro-cellulose) must be stabilised against decomposition.

Solid rocket propellants, when in compressed or cast form, must not contain any unintentional fissures or gas bubbles which dangerously affect their functioning.

4.3.7.6 Proof of Conformity

The procedures for the attestation of the conformity of explosives shall be either:

- EC-Type Examination and, at the choice of the manufacturer, either
  - type conformity (Module C);
  - Production Quality Assurance procedure (Module D);
  - Product Quality Assurance procedure (Module E);
  - Product Verification (Module F); or
- Unit Verification (Module G).

4.3.7.6.1 CE Marking

CE Marking has to be affixed so that it is visible, easily legible and indelible on the explosives themselves or, if this is not possible, on an identification plate that is attached to the explosive or, in the last resort, on the packaging.

The identification plate must be designed to prevent it being reused.

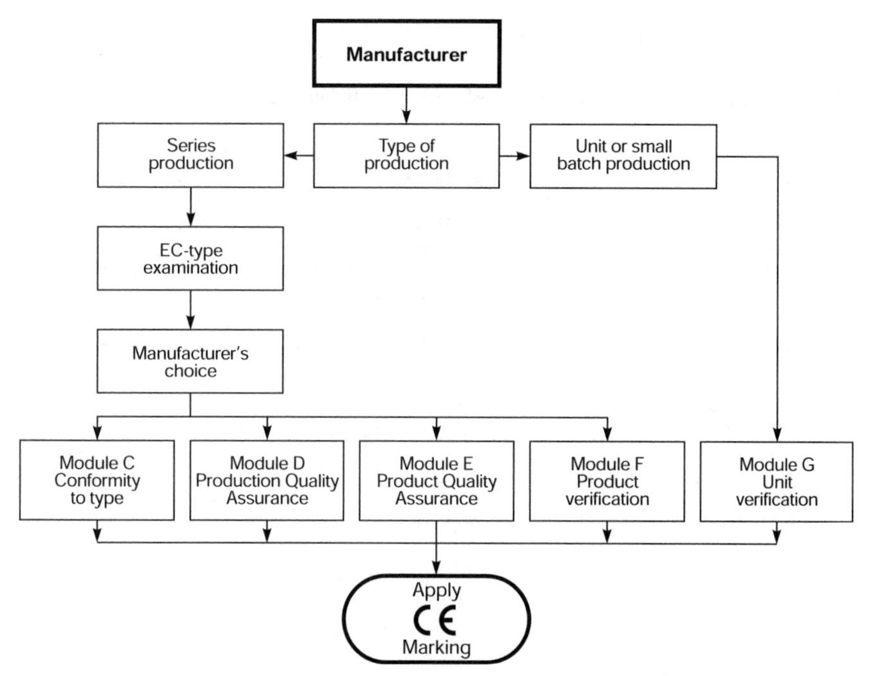

Figure 4.10 Conformity Procedure (93/15/EEC)

Annex	*Title*	*Yes/No*
	CE Marking	Yes
	EC Declaration of Conformity	Yes
	EC-Type Examination	Yes
	Product Quality Assurance	Yes
	Production Quality Assurance	Yes
	Product Verification	Yes
	Technical File	Yes
	Unit Verification	Yes
	Conformity to Type	Yes

Table 4.14 Proof of Conformity (93/15/EEC)

4.4 Other Directives

4.4.1 Simple Pressure Vessels (87/404/EEC)

Directive:	**87/404/EEC Dated: 25 Jun 87**
Short name	Simple Pressure Vessels or SPV
Description	Safety requirements for pressure vessels containing air or nitrogen
Amendments	90/488/EEC (provisions relating to the transitional period) 93/68/EEC (CE Marking)
Repealed Directives	
Member States Implementation Date	1 Jan 90 Implemented in the UK by The Simple Pressure Vessels (Safety) Regulations 1991 (SI 1991:2749)
Transitional Period	Up to 1 Jul 92, Pressure Vessels conforming to the rules in force in their territories before the date of application of the Directive are permitted to be placed on the market
Date of entry into force	1 Jul 90

4.4.1.1 Introduction

This Directive applies to simple pressure vessels manufactured in series. (i.e. any simple welded vessel subject to an internal gauge pressure greater than 0.4 bar, which is intended to contain air or nitrogen and which is not intended to be fired).

4.4.1.2 Structure

The Directive includes the Annexes Listed in Table 4.15.

4.4.1.3 Objectives

To ensure an appropriate level of safety throughout the EU for the uses of simple pressure vessels.

Annex	*Title*
I	Essential Requirements
II–1	EC Marks and Inscriptions
II–2	Instructions
II–3	Design and manufacturing schedules
II–4	Definitions and symbols
III	Minimum Criteria to be taken into account by Member States when appointing Inspection Bodies

Table 4.15 Annexes to Directive 87/404/EEC

4.4.1.4 Exclusions

The Directive does not apply to the following vessels:

- vessels specifically designed for nuclear use, failure of which may cause an emission of radioactivity;
- vessels specifically intended for installation in or on the propulsion of ships and aircraft;
- fire extinguishers.

4.4.1.5 Essential Requirements

The Directive states that:

- the parts and assemblies contributing to the strength of the vessel shall be made of either non-alloy quality steel or non-alloy aluminium or non-age-hardening aluminium alloy;
- the maximum working pressure of the vessel shall not exceed 30 bar and the product of that pressure and the capacity of the vessel (PS.V) shall not exceed 10,000 bar/litre;
- the minimum working temperature must be lower than –50°C and a maximum of 300 for steel and 100 for aluminium or aluminium alloy.

The Essential Safety Requirements also detail:

- the type of material to be used (i.e. pressurised parts, specifically for steel and/or aluminium vessels), welding materials and accessories and non-pressurised parts;
- the construction of the parts and assemblies contributing to the strength of the vessel under pressure;

- the cylindrical and cross-sectional construction of the vessel;
- vessel design (i.e. minimum/maximum working temperature, maximum working pressure, wall thickness, etc.);
- the manufacturing process (preparation of component parts, welds on pressurised parts);
- placing in service of the vessels (e.g. mandatory requirement for adequate maintenance and installation instructions).

4.4.1.6 Proof of Conformity

Vessels with a PS and V product less than 50 bar/litre (which are considered as representing only a minor hazard) do not have any obligation to meet the Essential Requirements contained in the Directive. The manufacturer is, however, required to ensure that the vessel is manufactured in accordance with sound engineering practice and (with the exception of the CE Marking) bears markings showing the maximum working pressure, etc.

Vessels with a PS and V product in excess of 50 bar/litre must satisfy the essential safety requirements. Certification of these vessels is as follows:

- if manufactured in accordance with the harmonised standards, the manufacturer can choose either to:
  - draw up a certificate of adequacy attesting that the schedule is satisfactory; or
  - submit a prototype vessel for the EC-Type Examination;
  - furthermore, if the product of PS and V exceeds 3000 bar/litre, then the vessel will require EC Verification. If PS and V is less than 3000 bar/litre then the manufacturer can choose either to issue an EC Declaration of Conformity or go along the EC Verification route.
- if the manufacture is not, or only partly, in accordance with the standards, then the manufacturer must submit a prototype of the production EC-Type Examination.

Title	*Yes/No*
CE Marking	Yes
EC Declaration of Conformity	Yes
EC Certificate of Conformity	Yes
EC-Type Examination	Yes
Technical File	Yes

Table 4.16 Proof of Conformity (87/404/EEC)

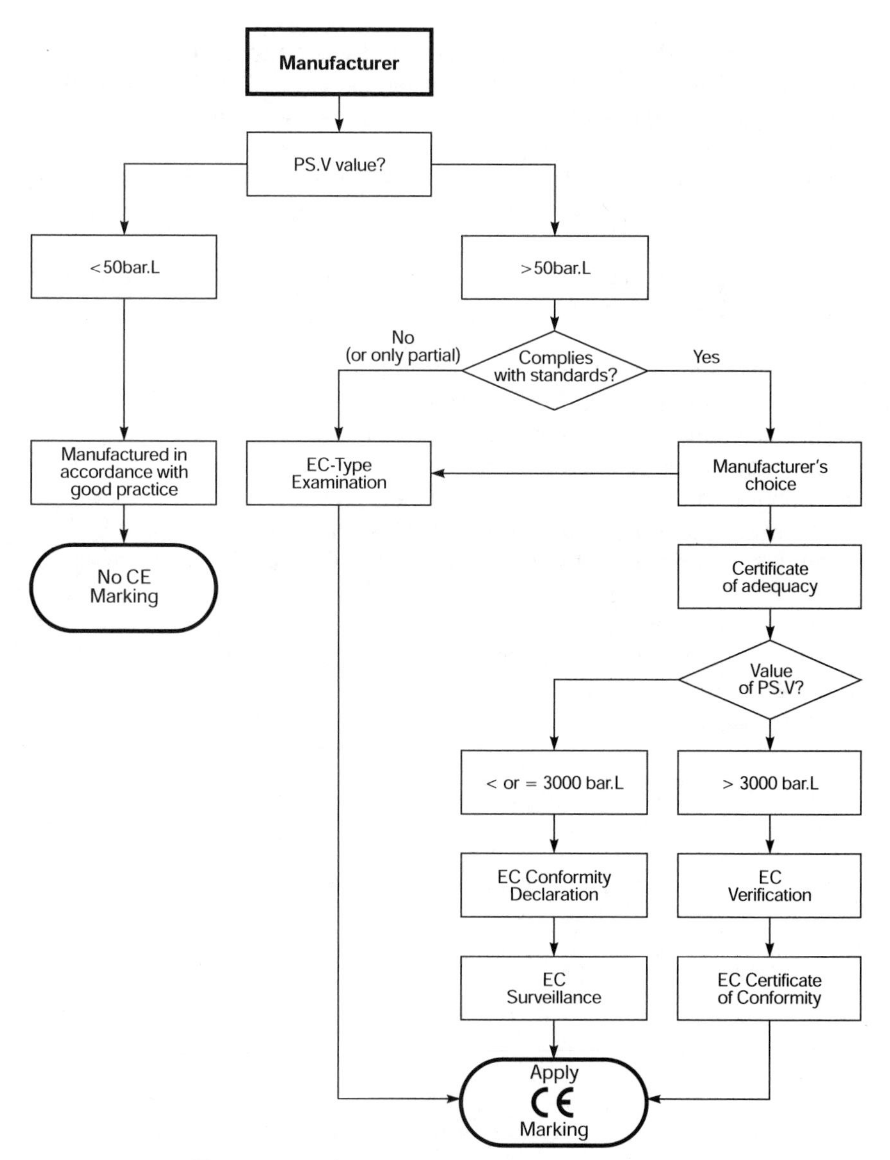

Figure 4.11 Conformity Procedure (87/404/EEC)

4.4.1.7 Other Relevant Information

A new Directive for Pressure Equipment is anticipated soon.

4.4.1.7.1 CE Marking

Whilst simple designs do not need to display the date of affixation, complete signs should indicate the year and shall consist of the symbol CE plus the last two digits of the year in which the mark was affixed.

Note: In the CE Amending Directive (93/68/EEC), the requirement to include the number of the body carrying out the EC-Type Examination was removed.

The vessel or data plate must bear the CE Marking together with at least the following information:

- the maximum working pressure PS in bar;
- the maximum working temperature T_{max} in °C;
- the minimum working temperature T_{min} in °C;
- the capacity of the vessel V in l;
- the name or mark of the manufacturer;
- the type and serial or batch identification of the vessel;
- the last two digits of the year in which the CE Marking was affixed.

Where a data plate is used, it must be so designed that it cannot be re-used and must include a vacant space to enable other information to be provided.

4.4.2 Pressure Equipment (97/23/EC)

Directive:	**97/23/EC** **Dated: 29 May 97**
Short name	Pressure Equipment Directive or PED
Description	Safety of plant and installations carrying pressures above 0.5 bar
Amendments	Nil
Repealed Directives	Nil
Member States Implementation Date	29 Nov 99 No additional UK legislation required for implementation although modification to some existing legislation is being considered
Transitional Period	Pressure equipment and assemblies which comply with the regulations in force in their territory at the date of application of this Directive until 29 May 2002 It has been agreed that both Directives (i.e. 87/404/EEC and 97/23/EC) are to be reviewed during 2003 to see if they can be amalgamated
Date of entry into force	28 May 2002

4.4.2.1 Introduction

The Pressure Equipment Directive (PED) is the second of a series of Directives (the first being 87/404/EEC, see 4.4.1) that are intended to replace the Framework Directive 76/767/EEC of 27 Jul 76.

97/23/EC mainly relates to equipment composed of several pieces of pressure equipment that have been assembled to constitute an integrated and functional whole. This equipment may range from simple assemblies such as pressure cookers to complex assemblies such as watertube boilers.

Note: This Directive applies **only** to equipment subject to a maximum allowable pressure PS exceeding 0.5 bar.

4.4.2.2 Structure

The Directive includes the Annexes Listed in Table 4.17.

Annex	*Title*
I	Essential safety requirements
II	Conformity assessment tables
III	Conformity assessment procedures Module A (Internal Production Control) Module B (EC-Type Examination) Module B1 (EC-design examination) Module C1 (Conformity to Type) Module D (Production Quality Assurance) Module D1 (Production Quality Assurance) Module E (Product Quality Assurance) Module E1 (Product Quality Assurance) Module F (Product Verification) Module G (EC Unit Verification) Module H (Full Quality Assurance) Module H1 (Full Quality Assurance with design and special surveillance of the final assessment)
IV	Minimum criteria to be met when designating the Notified Bodies and recognised third-party organisations referred
V	Criteria to be met when authorising User Inspectorates
VI	CE Marking
VII	Declaration of Conformity

Table 4.17 Annexes to Directive 97/23/EC

4.4.2.3 Objectives

The scope of 97/23/EC is based on a general definition of the term 'pressure equipment' so as to allow for the technical development of products.

4.4.2.4 Exclusions

The following are excluded from the scope of this Directive:

- equipment subject to a pressure of less than 0.5 bar (e.g. vessels designed for the transport and distribution of drinks);
- the assembly of pressure equipment on the site and under the responsibility of the manufacturer or supplier;
- pipelines comprising piping or a system of piping designed for the conveyance of any fluid or substance to or from an installation (onshore or offshore);

- networks for the supply, distribution and discharge of water (e.g. headraces, penstocks pressure tunnels, pressure shafts for hydro-electric installations);
- equipment covered by Directive 87/404/EEC on simple pressure vessels;
- equipment covered by Council Directive 75/324/EEC on aerosol dispensers;
- equipment intended for the functioning of vehicles defined in:
  - 70/156/EEC for type-approval of motor vehicles and their trailers;
  - 74/150/EEC for type-approval of wheeled agricultural or forestry tractors;
  - 92/61/EEC relating to the type-approval of two or three-wheel motor vehicles.
- equipment covered by one of the following Directives:
  - 89/392/EEC relating to machinery;
  - 95/16/EC relating to lifts;
  - 73/23/EEC relating to electrical equipment designed for use within certain voltage limits;
  - 93/42/EEC concerning medical devices;
  - 90/396/EEC relating to appliances burning gaseous fuels;
  - 94/9/EC concerning equipment and protective systems intended for use in potentially explosive atmospheres;
  - equipment covered by Article 223 (1) (b) of the Treaty.
- items specifically designed for nuclear use;
- well-control equipment used in the petroleum, gas or geothermal exploration and extraction;
- casings and/or machinery for:
  - engines including turbines and internal combustion engines;
  - steam engines, gas/steam turbines, turbo-generators, compressors, pumps and actuating devices;
  - blast furnaces, hot-blast recuperators, dust extractors, blast-furnace exhaust-gas scrubbers, direct reducing cupolas, gas converters and pans for melting, re-melting, de-gassing and casting of steel and non-ferrous metals.
- enclosures for high-voltage electrical switchgear, control gear, transformers, and rotating machines;
- pressurised pipes for electrical power and telephone cables;
- ships, rockets, aircraft and mobile off-shore units;
- pressure equipment consisting of a flexible casing, e.g. tyres, air cushions, balls used for play, inflatable craft, and other similar pressure equipment;
- exhaust and inlet silencers;
- bottles or cans for carbonated drinks for final consumption;
- radiators and pipes in warm water heating systems.

4.4.2.5 Essential Requirements

The Essential (safety) Requirements have been subdivided into general and specific requirements. These requirements are contained in 10 pages of text and cover:

4.4.2.6 General

- **Design** – incorporation of safety coefficients, loading, strength (this section also goes into great detail about calculation and test methods), safe handling and operation, means of inspection (draining and venting), corrosion, wear, assemblies, safety accessories and external fire;
- **Manufacture** – preparation of component parts, non-destructive tests, heat treatment, tractability, final inspection, proof test, inspection of safety devices, making and labelling and operating instructions;
- **Materials** – sufficiently ductile and tough, chemically resistant to the fluid contained in the pressure equipment, sufficiently unaffected by aging.

4.4.2.7 Specific Pressure Equipment Requirements

This is a separate section and is split into:

- fixed or otherwise heated pressure equipment with a risk of overheating;
- piping;
- specific quantitative requirements for certain pressure equipment.

Title	*Yes/No*
CE Marking	Yes
EC Declaration of Conformity	Yes
EC-Type Examination	Yes
Product Quality Assurance	Yes
Production Quality Assurance	Yes
Full Quality Assurance	Yes
Product Verification	Yes
Technical File	Yes
Unit Verification	Yes
Internal Production Control	Yes
Conformity to Type	Yes

Table 4.18 Proof of Conformity (97/23/EC)

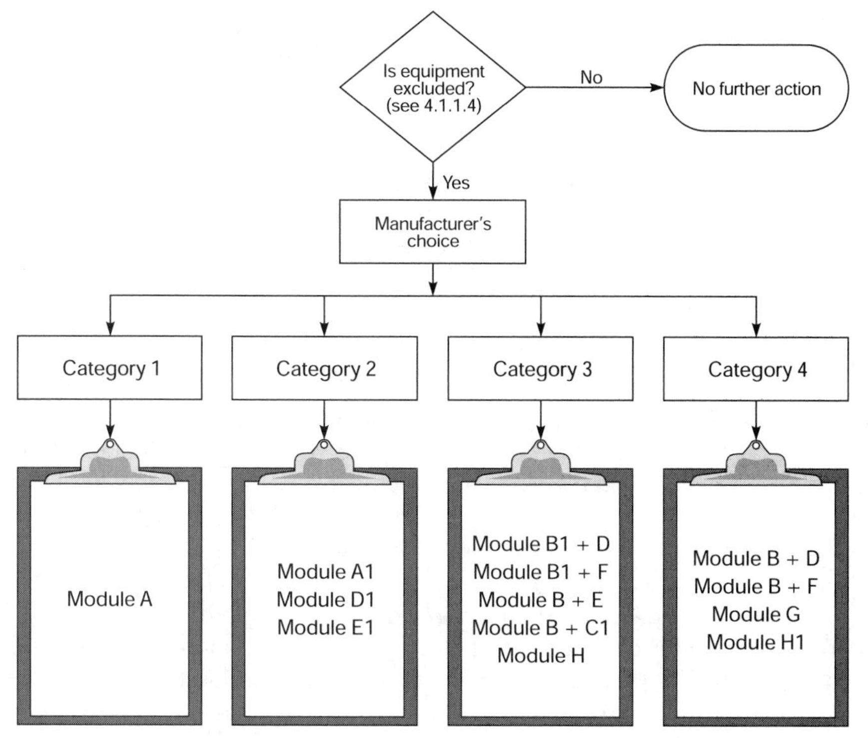

Figure 4.12 Conformity Procedure (97/23/EC)

4.4.2.8 Classification of Pressure Equipment

Pressure equipment is classified by category according to an ascending level of hazard and into two groups.

Group 1 comprises dangerous fluids defined as:

- explosive;
- extremely flammable;
- highly flammable;
- flammable (where the maximum allowable temperature is above flashpoint);
- very toxic;
- toxic;
- oxidizing.

Group 2 comprises all other fluids.

Notes

1 Where a vessel is composed of a number of chambers, it shall be classified in the highest category applicable to the individual chambers.

2 Where a chamber contains several fluids, classification shall be on the basis of the fluid which requires the highest category.

4.4.2.9 Conformity Assessment

At the choice of the manufacturer, pressure equipment is subjected to one of the conformity assessments shown against the category in which it is classified. The manufacturer may also choose to apply one of the procedures, which apply to a higher category, if available. The various categories are as follows:

Category 1	**Category II**	**Category III**	**Category IV**
Module A	Module A1	Module B1 + D	Module B + D
	Module D1	Module B1 + F	Module B + F
	Module E1	Module B + E	Module G
		Module B + C1	Module H1
		Module h	

4.4.2.10 Other Relevant Information

The Department of Trade and Industry (DTI) will prepare implementing regulations. The proposed regulations will be issued for public comment any time now.

The UK Health and Safety Executive/Council (HSE/HSC) are currently looking at the affect the PED has on existing legislation in Britain covering manufacture, assembly, installation and putting into service of pressure equipment. A Consultation Package on the proposed changes (which includes a revised ACOP) will be issued shortly.

4.4.2.10.1 Publications

DTI have issued a free guidance booklet in the Business in Europe – Product Standards series entitled *Pressure equipment*, which is available from DTI (see *Yellow Pages*).

4.4.2.10.2 CE Marking

It is not necessary for the CE Marking to be affixed to each individual item of pressure equipment making up an assembly.

In addition to the CE Marking, the following information must be provided:

For all pressure equipment:

- the name and address or other means of identification of the manufacturer and, where appropriate, of his authorised representative established within the EU;

- the year of manufacture;
- identification of the pressure equipment according to its nature, such as type, series or batch identification and serial number;
- essential maximum/minimum allowable limits.

Depending on the type of pressure equipment, further information necessary for safe installation, operation or use and, where applicable, maintenance and periodic inspection such as:

- the volume V of the pressure equipment in L;
- the nominal size for piping DN;
- the test pressure PT applied in bar and date;
- safety device set pressure in bar;
- output of the pressure equipment in kW;
- supply voltage in V (volts);
- intended use;
- filling ratio kg/l;
- maximum filling mass in kg;
- tare mass in kg;
- the product group.

4.4.3 Active Implantable Medical Devices (90/385/EEC)

Directive:	**90/385/EEC** **Dated: 20 Jun 90**
Short name	Active Implantable Medical Devices
Description	Safety and performance requirements for electrically operated medical devices intended for implantation in the human body
Amendments	93/68/EEC (CE Marking) 93/42/EEC (Medical Devices)
Repealed Directives	
Member States Implementation Date	31 Dec 92 Implemented in the UK by the Active Implantable Medical Devices Regulations 1992 (SI 1992:3146)
Transitional Period	ends 21 Dec 95
Date of entry into force	31 Dec 94

4.4.3.1 Introduction

To ensure that the health and safety of patients and users can be guaranteed and that the manufacture of medical devices can be regulated throughout the EU, Member States have agreed the necessity for a series of Directives regulating the safety and marketing of medical devices. These directives came into effect from January 1993.

This is the first of two main directives and covers all powered implants or partial implants that are left in the human body. Heart pacemakers are the most common example of powered implants. The Directive was implemented in the UK by the Active Implantable Medical Devices Regulations (SI 1995 No 1671 as amended) which came into effect on 1 January 1995.

4.4.3.1.1 Structure

The Directive includes the Annexes Listed in Table 4.19.

Annex	*Title*
I	Essential Requirements
II	EC Declaration of Conformity (Complete quality assurance system)
III	EC-Type Examination
IV	EC Verification
V	EC Declaration of Conformity (Assurance of production quality)
VI	Statement concerning devices intended for special purposes
VII	Clinical evaluation
VIII	Minimum criteria to be met when designating Inspection Bodies to be Notified
IX	CE Mark of Conformity

Table 4.19 Annexes to Directive 90/385/EEC

4.4.3.2 Objectives

To harmonise and improve the safety standards of active implantable medical devices which are intended to be partially or totally introduced, surgically or medically, into a human body or by medical intervention into a natural orifice and which rely for their functioning on an electrical source or source other than that directly generated by the human body or gravity.

4.4.3.3 Exclusions

Where an active implantable medical device is intended to administer a substance defined as a medicinal product it shall be subject to the requirements of Directive 65/65/EEC.

4.4.3.4 Essential Requirements

For Active Implantable Medical Devices the Essential Requirements are, quite understandably, very complex. Full details are available in Annex 1 of the Directive and consist of:

General requirements

- Active implantable medical devices must provide patients, users and other persons with a high level of protection and be capable of achieving their intended level of performance when implanted in human beings.
- Their use must not compromise the clinical condition or the safety of patients.
- They must not present any risk to the persons implanting them.
- They must achieve the performance intended by the manufacturer.
- Any side effects or undesirable conditions must constitute acceptable risks when weighed against the performances intended.

Requirements regarding design and construction

- They must be designed, manufactured and packed in a non-reusable pack;
- They must be designed to minimise the risk of physical injury, risk connected with medical treatment;
- There must be choice of materials (particularly with regard to toxicity aspects);
- There must be mutual compatibility between the materials used and biological tissues, cells and body fluids;
- They must comply with labelling requirements for the sterile pack and on the sales packaging.

4.4.3.5 Proof of Conformity

For medical devices (other than those which are custom-made or intended for clinical investigations) the manufacturer shall (at his own choice) either;

- follow the procedure relating to the EC Declaration of Conformity; or
- follow the procedure relating to EC-Type Examination coupled with either EC Verification or the procedure relating to the EC Declaration of Conformity to type.

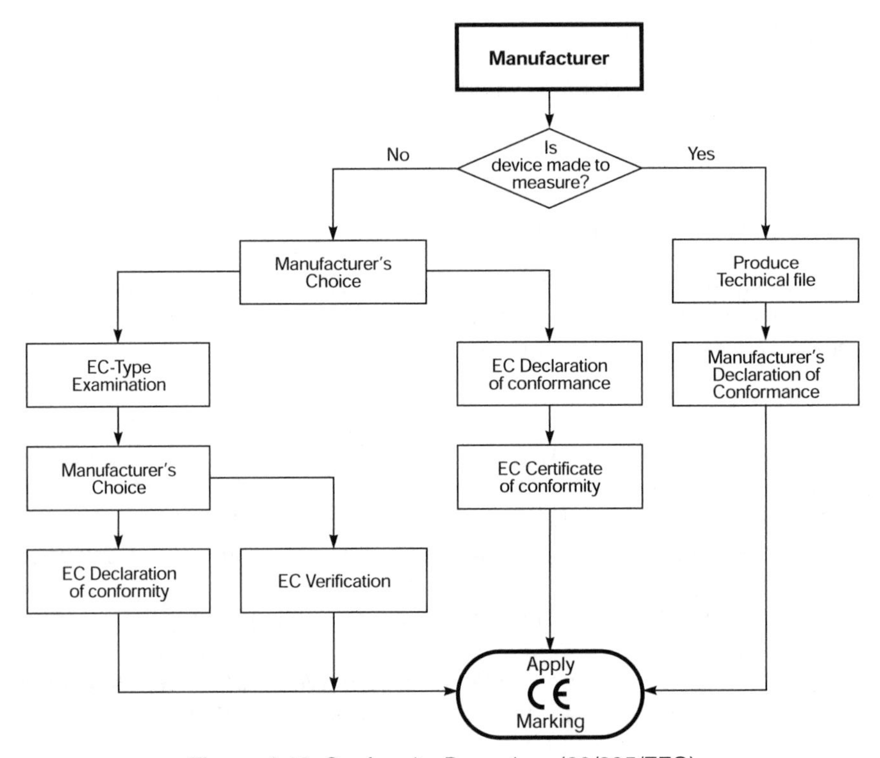

Figure 4.13 Conformity Procedure (90/385/EEC)

Title	*Yes/No*
CE Marking	Yes
EC Declaration of Conformity	Yes
EC-Type Examination	Yes
Product Verification	Yes
Technical File	Yes

Table 4.20 Proof of Conformity (98/385/EC)

In the case of custom-made devices, the manufacturer must draw up a declaration (details of which are provided in Annex 6 to the Directive) containing a general description of the device, how it is to be used and how it should be tested, etc. This declaration must be submitted to the Competent Authorities of the Member State in question at least 60 days before the commencement of the investigations.

4.4.3.5.1 Publications

- Application Guide 'Guidelines to the demarcation between Directives 9/385/EEC, 93/42/EEC and 65/65/EEC';
- Application Guide 'Guidelines on a Medical Vigilance System'.

4.4.3.5.2 CE Marking

CE Marking must appear in a visible, legible and indelible form on the sterile pack and, where appropriate, on the sales packaging, if any, and on the instruction leaflet. It must be accompanied by the logo of the Notified Body. The minimum vertical dimension of 5 mm may be waived for small-scale devices.

The following additional information regarding the affixing of CE Marking was made by the CE Amending Directive (93/68/EEC). 'CE Marking must now be followed by the identification number of the Notified Body instead of the log'. There are also changes to the verification unit and new administration procedures.

4.4.4 Medical Devices (93/42/EEC)

Directive:	**93/42/EEC Dated: 4 Jun 93**
Short name	Medical Devices or MDD
Description	Safety of all medical equipment not covered by directives on in-vitro fertilisation or active implantable devices
Amendments	Directive 84/539/EEC (electro-medical equipment used in human or veterinary medicine) requires amendment
Repealed Directives	76/764/EEC (Clinical mercury-in-glass, maximum reading thermometers)
Member States Implementation Date	1 Jul 94 Implemented in the UK by the Medical Devices Regulations 1994 (SI 1994:3017)
Transitional Period	Devices that conform to existing rules are granted a transitional period until 13 Jun 98. Devices covered by Directive 76/764/EEC may be placed on the market up until 30 Jun 04
Date of entry into force	1 Jan 95

4.4.4.1 Introduction

This is the second Directive from the series of Directives regulating the safety and marketing of medical devices throughout the European Union.

This Directive (which is based largely on the provisions of the first Directive, 90/385/EEC, see 4.4.3) covers most of the medical devices, ranging from, for example, first aid bandages, tongue depressors, hip prostheses, X-ray equipment, ECG and heart valves. The Directive was implemented in the UK by the Medical Devices Regulations (SI 1994 No 3017).

4.4.4.2 Structure

The Directive includes the Annexes Listed in Table 4.21.

Annex	*Title*
Annex I	Essential Requirements
Annex II	EC Declaration of Conformity (Full Quality Assurance System)
Annex III	EC-Type Examination
Annex IV	EC Verification
Annex V	EC Declaration of Conformity (Production quality assurance)
Annex VI	EC Declaration of Conformity (Product quality assurance)
Annex VII	EC Declaration of Conformity
Annex VIII	Statement concerning Devices for Special Purposes
Annex IX	Classification Criteria
Annex X	Clinical Evaluation
Annex XI	CE Marking of Conformity
Annex XII	Criteria to be met for the designation of Notified Bodies

Table 4.21 Annexes to Directive 93/42/EEC

4.4.4.3 Objectives

The Medical Devices Directive (i.e. 93/42/EEC) is an all-encompassing document covering the manufacture of any medical device or material used either temporarily or permanently in, or on, human beings.

The main aim of the MDD is to ensure that all medical devices that are placed on the market do not compromise the safety and health of patients (and users) when properly installed, maintained and used in accordance with their intended purpose. Patients, users and third parties should, therefore, be provided with a high level of protection and the devices should attain the performance levels claimed by the manufacturer.

4.4.4.4 Exclusions

This Directive does not apply to:

- in-vitro diagnostic devices covered by Directive 98/79/EEC;
- active implantable devices covered by Directive 90/385/EEC;
- medicinal products covered by Directive 65/65/EEC;
- cosmetic products covered by Directive 76/768/EEC;
- human blood, human blood products, human plasma or blood cells of human origin;
- transplants or tissues or cells of human origin;
- transplants or tissues or cells of animal origin;
- personal protective equipment covered by Directive 89/686/EEC.

This Directive does not affect the application of Directive 80/836/Euratom, nor of Directive 84/466/Euratom.

Note: By definition, the placing on the market of a **medical device** is, as a general rule, governed by the Medical Devices Directive. On the other hand, the placing on the market of a **medicinal product** is governed by Directive 65/65/EEC. If, however, the device **and** the medicinal product form a single integral unit which is intended exclusively for use in the given combination and which is not reusable, then that single unit product shall be governed by Directive 65/65/EEC.

4.4.4.5 Essential Requirements

The Essential Requirements for Medical Devices fall under two headings, 'General' and 'Design and Construction'.

4.4.4.5.1 General

Medical devices shall be designed and manufactured so that when they are used under normal conditions (i.e. under which they were originally intended), they will not compromise the clinical condition, safety, or health of users and/or patients.

The manufacturer shall:

- eliminate risks as far as possible (i.e. provide an inherently safe design and construction);
- take adequate protection measures (including alarms if necessary) for any risks that cannot be completely eliminated;
- inform users of any residual risks due to any shortcomings of the protection measures that they have adopted;
- any side effect caused by a medical device shall not constitute an unacceptable risk to the (original) performances intended.

4.4.4.5.2 Design and Construction Requirements

The requirements shown in the Directive cover six pages and are subdivided as follows:

- chemical, physical and biological properties;
- infection and microbial contamination;
- construction and environmental properties;
- devices with a measuring function;
- protection against radiation (intended and unintended);
- requirements for medical devices connected to or equipped with an energy source (protection against electrical risks, mechanical and thermal risks and risks posed to the patient by energy supplies or substances);
- information to be supplied by the manufacturer (labels, cleaning and sterilisation instructions, instructions concerning patient safety).

4.4.4.6 Proof of Conformity

Devices covered by the Directive are grouped into four Classes as follows:

- Class I – generally regarded as low risk;
- Class IIa – regarded as medium risk;
- Class IIb – regarded as medium risk;
- Class III – generally regarded as high risk.

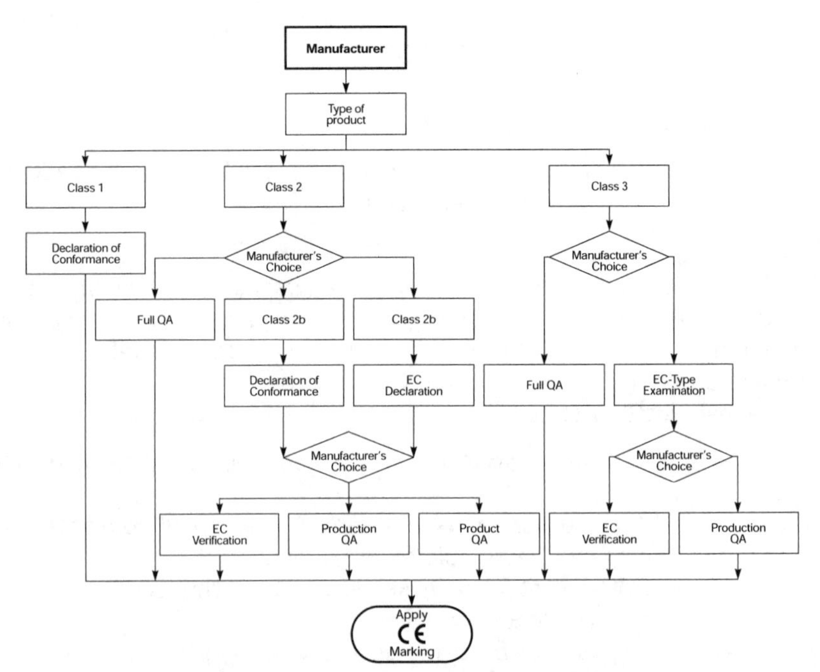

Figure 4.14 Conformity Procedure (93/42/EEC)

Title	*Yes/No*
CE Marking	Yes
EC Declaration of Conformity	Yes
EC-Type Examination	Yes
Product Quality Assurance	Yes
Production Quality Assurance	Yes
Full Quality Assurance	Yes
Product Verification	Yes
Technical File	Yes

Table 4.22 Proof of Conformity (93/42/EEC)

The classification of a device depends on a whole number of variables such as the type of device, where it is to be located, how long it is to be used for, whether it is implanted, etc.

4.4.4.6.1 Class I

The manufacturer of all devices falling within Class I (other than devices which are custom-made or intended for clinical investigations), shall draw up the EC Declaration of Conformity before placing the device on the market.

4.4.4.6.2 Class IIa

Manufacturers of medical devices (other than devices that are custom-made or intended for clinical investigations falling within Class IIa), shall follow the procedure relating to EC Declaration of Conformity coupled with either one of the following:

- the procedure relating to the EC Verification set;
- the procedure relating to the EC Declaration of Conformity (Production Quality Assurance);
- the procedure relating to the EC Declaration of Conformity (Product Quality Assurance).

Alternatively, the manufacturer can apply Full Quality Assurance.

4.4.4.6.3 Class IIb

Manufacturers of devices falling within Class IIb (other than devices which are custom-made or intended for clinical investigations) need to:

- follow the procedure relating to the EC-Type Examination (Full Quality Assurance) – especially in the case of custom-made devices;

- follow the procedure relating to the EC-Type Examination coupled with:
  - the procedure relating to the EC verification or
  - the procedure relating to the EC Declaration of Conformity (Production Quality Assurance) or
  - the procedure relating to the EC Declaration of Conformity (Product Quality Assurance).

4.4.4.6.4 Class III

Manufacturers of devices falling within Class III (other than devices which are custom-made or intended for clinical investigations) shall either:

- follow the procedure relating to the EC Declaration of Conformity (Full Quality Assurance); or
- follow the procedure relating to the EC-Type Examination coupled with either:
  - the procedure relating to the EC verification; or
  - the procedure relating to the EC Declaration of Conformity (Production Quality Assurance).

4.4.4.6.5 Custom-Made and Clinical Investigation Devices

The conformity assessment requirements for devices that are custom-made or intended for clinical investigations are detailed in MDD Annex VIII.

4.4.4.6.6 Particular Requirements for System or Procedure Packs

Any person who places medical devices on the market as a system or procedure pack shall draw up a declaration by which he states that:

- he has verified the mutual compatibility of the devices in accordance to the manufacturers' instructions;
- he has packaged the system or procedure pack and supplied relevant information to users.

The whole activity is subjected to appropriate methods of internal control and inspection.

4.4.4.7 Other Relevant Information

4.4.4.7.1 Review

The Commission shall, no later than five years from the date of implementation of this Directive, submit a report to the Council on the operation of the provisions.

4.4.4.7.2 Clinical Investigation

Manufacturers of devices that are intended for clinical investigations shall inform the Notified Body and follow the procedure referred to in Annex VIII of the Directive.

4.4.4.7.3 Implantable and Long-Term Invasive Devices

Manufacturers of devices falling within Class III **and** which are implantable and long-term invasive devices falling within Class IIa or IIb, may commence the relevant clinical investigation at the end of a period of 60 days following notification to a Notified Body, unless the Competent Authorities have said otherwise.

Note: This clinical evaluation is aimed at identifying any undesirable side-effects and to verify that the device meets the general requirements detailed in Annex I of the Directive.

4.4.4.7.4 HIV Virus

Medical devices used for protection against the HIV must ensure that there is a high protection level and the design and a Notified Body has verified manufacture of these products.

4.4.4.7.5 EMC

Medical devices shall conform to the requirements of Council Directive 89/336/EEC for all aspects of electromagnetic compatibility and any MDD system specific requirements that are not covered by this generic standard should be stated.

4.4.4.7.6 Ionising Radiation

Manufacturers' specifications shall pay due attention to the requirements regarding the design and manufacture of devices emitting ionising radiation.

Conformity Assessment Procedures are laid down in Annexes to the Directive and vary for:

- each Class;
- the generic type of device (i.e. non-invasive, invasive, active or those with higher than expected risk).

4.4.4.7.7 Publications

- Application Guide 'Guidelines relating to the demarcation between Directives 90/385/EEC, 93/42/EEC and 65/65/EEC'.
- Application Guide 'Guidelines on a Medical Devices Vigilance System'.
- 'Getting a Good Deal in Europe'. This guide may be obtained from the European Community Section of the DTI's Deregulation Unit (Tel: 0171 215 6394).

4.4.4.7.8 CE Marking

All medical devices (other than those that are custom-made or used for clinical investigations) shall bear the CE Marking to indicate conformity with the provisions of the Directive and to enable them to move freely within the EU.

The minimum vertical dimension of 5 mm may be waived for small-scale devices.

Medical Devices which do not need to have the CE Marking

For the purpose of the MDD Directive, it has been agreed that the following devices do not need to have the CE Marking:

- devices intended for clinical investigation being made available to medical practitioners or authorised persons;
- custom-made devices being placed on the market or being put into service.

4.4.5 Medical Devices – In Vitro Diagnostics (COM(95)130)

Directive:	**Not yet allocated** **Dated:**
Short name	In Vitro Diagnostics (IVD)
Description	
Amendments	
Repealed Directives	
Member States Implementation Date	
Transitional Period	
Date of entry into force	

4.4.5.1 Introduction

This will be the third of a series of Directives regulating the safety and marketing of medical devices throughout the European Union. At the time of writing, the Directive (98/79/EC) was being revised and from information received it is understood that a new version will be distributed in 2000.

4.4.5.2 Structure

It is understood that the layout will be very similar to the Medical Devices Directive (see 4.4.4).

4.4.5.3 Objectives

The Directive will cover any medical device which is a reagent, calibrator, control material, kit, instrument, apparatus, equipment or system intended for use in vitro for the examination of specimens, including blood and tissue donations, derived from the human body, for the purpose of providing information concerning a physiological state, state of health or disease or congenital abnormality or to determine the safety and compatibility with potential recipients.

Examples of in vitro diagnostic devices are blood grouping reagents, pregnancy test kits and hepatitis B test kits. The Directive will not be implemented in the UK until June 2000 and will have a 3½ year transitional period.

4.4.5.4 Exclusions

The following will be excluded from this Directive:

- reagents produced within user laboratories and which are not subject to commercial transactions and products intended for general laboratory use, unless they are specifically intended to be used for in vitro diagnostic examination;
- devices manufactured and used only within the same institution and on the premises of their manufacture.

4.4.5.5 Essential Requirements

It is, however, understood that the Essential Requirements will be very similar to the Medical Devices Directive (see 4.4.4.5).

4.4.5.6 Proof of Conformity

Most in vitro medical devices do not constitute a direct risk to patients and are used by competently trained professionals with results often confirmed by other means. For these, conformity assessment can usually be carried out under the sole responsibility of the manufacturer under the EC Declaration of Conformity procedure.

For other defined devices where correct performance is essential to medical practice and where failure can cause a serious risk to health, there must be a third-party conformity assessment. These apply to:

- reagents and reagent products for blood grouping (A, B, 0 and Rho/D);
- the detection in humans of markers of HIV infection and hepatitis B and C.

4.4.5.7 Other Relevant Information

A specimen receptacle specifically intended to contain a specimen for the purpose of in vitro diagnostic examination is classed as a device. These devices will usually be used by medical laboratories, doctors, pathologists and medical scientists although some may be used by the patients themselves, e.g. pregnancy tests.

4.4.6 Non Automatic Weighing Instruments (90/384/EEC)

Directive:	**90/384/EEC Dated: 20 Jun 90**
Short name	Non Automatic Weighing Instruments
Description	Performance and calibration procedures for commercial weigh-scales
Amendments	93/68/EEC (CE Marking)
Repealed Directives	73/360/EEC, except for the transitional requirements
Member States Implementation Date	1 Jan 93 Implemented in the UK by the Weighing Instruments (EEC Requirements) Regulations 1992
Transitional Period	Until 1 Jul 02, all Instruments that conformed to the rules in force before the application date
Date of entry into force	1 Jan 93

4.4.6.1 Introduction

This Directive applies to all non-automatic weighing instruments (i.e. a weighing instrument that does not need the intervention of an operator during weighing).

It is 'total in application' meaning that no national provisions dealing with the application field of the directive may exist any more, other than those transposing the directive.

4.4.6.2 Structure

The Directive includes the Annexes listed in Table 4.23.

Annex	*Title*
I	Essential Requirements
II–1	EC-Type Examination
II–2	EC Declaration of Type Conformity (guarantee of production quality)
II–3	EC Verification
II–4	EC Unit Verification
II–5	Common provisions
III	Design documentation
IV	CE Marking
V	Minimum criteria for Notified Bodies

Table 4.23 Annexes to Directive 90/384/EEC

4.4.6.3 Objectives

To protect the public against incorrect weighing operations made from non-automatic weighing instruments.

To set the essential (metrological and performance) requirements that are necessary to guarantee effective protection for consumers and to lay down certification rules and procedures.

4.4.6.4 Exclusions

- coin counters, pill counters, parts counters, etc.;
- sports and sporting records;
- domestic use (kitchen, bathroom, etc.);
- geological surveys;
- veterinary medicine;
- medical practice, except for the weighing of live patients;
- goods inwards inspection, etc. (checking scales);

- weighing of goods for customer information only and not for the final determination of mass (not to be confused with self-service scales that are used for the final determination of mass);
- also the use of instruments by:
  - metrological enforcement agents;
  - metrological experts (e.g. national metrology laboratories, etc.);
  - government or public authority laboratories, forensic laboratories, etc.

4.4.6.5 Essential Requirements

These are the technical and metrological performance requirements for the instrument and include the maximum permitted error. The Essential Requirements cover seven pages of the Directive and are split into 'Metrological' and 'Design and Construction' Requirements.

4.4.6.5.1 Metrological Requirements

- units of mass – permitted ranges (i.e. SI, Imperial and non-SI units) defined in Directive 80/181/EEC as last amended by Directive 85/1/EEC;
- accuracy classes – e.g. scale intervals, verification scale defined for special, high, medium, ordinary classes;
- weighing ranges – requirements depending on the classification (i.e. instruments with one weighing range, multiple weighing ranges or multi-interval instruments);
- accuracy – error of indication, maximum permissible errors, etc.;
- weighing results – shall be repeatable and reproducible, sufficiently insensitive to changes in the position of the load on the load receptor;
- the instrument shall react to small variations in the load;
- influence quantities and time – insensitive to the degree of tilting, mains or battery supply fluctuations, relative humidity levels.

4.4.6.5.2 Design and Construction

- general requirements – the value of the mass must be indicated, shall be unaffected by disturbances, shall automatically detect and indicate faults, shall have no characteristics likely to facilitate fraudulent use;
- indication of weighing results and other weight values – shall be accurate, unambiguous and non-misleading;
- printing of weighing results and other weight values – clear, legible, non-erasable and durable;
- levelling – when appropriate, instruments shall be fitted with a levelling device and a level indicator;
- zeroing – if instruments are equipped with zeroing devices, they must be accurate and not cause incorrect measuring results;
- tare devices and pre-set tare devices – shall enable correct determination of the calculated net value;

- additional requirements for instruments for direct sales to the public – shall show all essential information about the weighing operation and, in the case of price-indicating instruments, shall clearly show the customer the price calculation of the product to be purchased;
- price labelling instruments – requirements, applicability, minimum capacity.

4.4.6.6 Proof of Conformity

The conformity of instruments to the Essential Requirements is a manufacturer's choice between:

- completing the EC-Type Examination followed by either the EC Declaration of Type Conformity (guarantee of production quality), or by the EC verification procedure;
- EC Unit Verification.

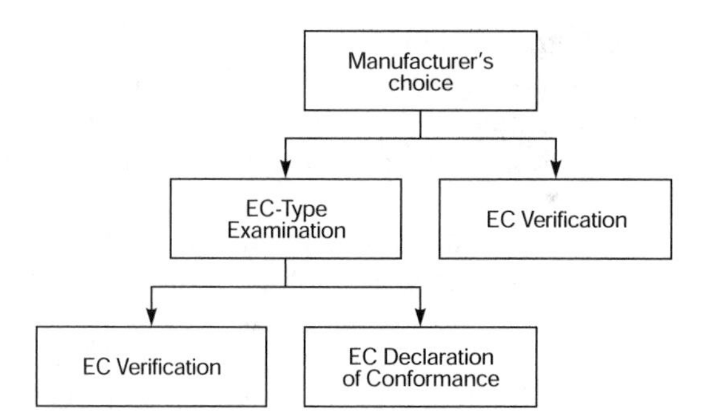

Figure 4.15 Conformity Procedure (90/384/EEC)

Title	*Yes/No*
CE Marking	
EC Declaration of Conformity	Yes
EC Certificate of Conformity	
EC-Type Examination	Yes
Product Verification	Yes
Technical File	Yes
Unit Verification	Yes

Table 4.24 Proof of Conformity (90/384/EEC)

Note: EC-Type Examination is not compulsory for instruments which do not use electronic devices and whose load-measuring device does not use a spring to balance the load.

4.4.6.6.1 Harmonised Standards

A harmonised European standard for non-automatic weighing instruments, EN 45501:1992/AC:1993, has been adopted by CEN/CENELEC. This standard not only sets down technical solutions to the Essential Requirements but also the tests that can be used to establish conformity with the performance requirements.

4.4.6.6.2 Publications

Application Guide – Document WELMEC 5. This document provides explanation and interpretation in respect of the provisions contained in Council Directive 90/384/EEC and is available from WELMEC, National Weights and Measures Laboratory, Stanton Avenue, Teddington, Middlesex TW11 0JZ, UK, Tel. +44 181 943 7298, Fax +44 181 943 7270.

4.4.6.6.3 CE Marking

Devices which are connected to or included as part of an instrument with CE Marking and which have not been subject to conformity assessment must bear the red symbol of restricted use with the capital letter M in black crossed by an X.

The CE Marking, if accompanied by the green metrology sticker (which is a supplementary, mandatory, metrology marking), signifies that the instrument carrying these two markings satisfies the Essential Requirements of the Directive.

As indicated in the CE amending Directive (93/68/EEC), the year must be indicated on the equipment but is no longer required for the CE Marking.

4.4.7 Gas Appliances (90/396/EEC)

Directive:	**90/396/EEC Dated: 29Jun 90**
Short name	Gas Appliances
Description	Safety of all appliances burning gaseous fuels
Amendments	93/68/EEC (CE Marking)
Repealed Directives	84/530/EEC (Framework Directive on gas appliances) – as amended by 86/312/EEC 84/531/EEC (gas water heaters) – as amended by 88/665/EEC
Member States Implementation Date	1 Jul 91 Implemented in the UK by the Gas Appliances (Safety) Regulations 1992 (SI 1992:711)
Transitional Period	Until 31 Dec 95, all appliances that conformed to the rules in force before 1 Jan 92
Date of entry into force	1 Jan 92

4.4.7.1 Introduction

This Directive for gas appliances is an updated amalgamation of the framework Directive on gas appliances (84/530/EEC – amended by 86/312/EEC) and the Directive on gas water heaters (84/531/EEC – amended by Directive 88/665/EEC) and applies to:

- all appliances burning gaseous fuels used for cooking, heating, hot water production, refrigeration, lighting or washing and having, where applicable, a normal water temperature not exceeding 105°C;
- safety devices, controlling devices or regulating devices and sub-assemblies, other than forced draught burners.

4.4.7.2 Structure

The Directive includes the Annexes listed in Table 4.25.

Annex	*Title*
I	Essential Requirements
II	Procedure for Certification of Conformity
II–1	EC-Type Examination
II–2	EC Declaration of Conformity to Type
II–3	EC Declaration of Conformity to Type (guarantee of production quality)
II–4	EC Declaration of Conformity to Type (guarantee of product quality)
II–5	EC Verification
III–6	EC Verification by Unit
III	CE Marking and Inscriptions
IV	Design Documentation
	Part B – Deadlines for Transposition into National Law
V	Minimum criteria for the assessment of Notified Bodies

Table 4.25 Annexes to Directive 90/396/EEC

4.4.7.3 Objectives

To ensure harmonisation of the characteristics that are necessary to meet the Essential Requirements for the health and safety of persons, domestic animals and property with regards to hazards generated by appliances burning gaseous fuels.

4.4.7.4 Exclusions

Appliances specifically designed for use in industrial processes carried out on industrial premises.

4.4.7.5 Essential Requirements

The Essential Requirements cover ‘General Conditions’, ‘Materials’ and ‘Design and Construction’.

4.4.7.5.1 General Conditions

- Appliances must be so designed and built as to operate safely and present no danger to persons, domestic animals or property when normally used.

- All appliances must be accompanied by technical instructions intended for the installer (e.g. installation, adjustment and servicing). In particular, the instructions must specify:
  - type of gas used;
  - gas supply pressure;
  - flow of fresh air required;
  - conditions for the dispersal of combustion products.
- All appliances must be accompanied by technical instructions intended for the user (e.g. use and servicing);
- All appliances must bear appropriate warning notices.

4.4.7.5.2 Materials

Materials must:

- be appropriate for their intended purpose;
- be capable of withstanding the envisaged technical, chemical and thermal conditions;
- guarantee the properties of materials.

4.4.7.5.3 Design and Construction

General

- appliances must be so constructed that, when used normally, no instability, distortion, breakage or wear likely to impair their safety can occur;
- condensation produced at the start-up and/or during use must not affect the safety of appliances;
- the risk of explosion in the event of an external fire must be minimised;
- the possibility of water and/or air penetration into the gas circuit shall be eliminated;
- auxiliary energy failure (or fluctuation) shall not lead to an unsafe situation;
- the design of the appliance shall guard against electrical hazards;
- all pressurised parts shall be capable of withstanding the envisaged mechanical and thermal stresses;
- failure of a safety, controlling or regulating devices shall not lead to an unsafe situation;
- the functioning of the safety devices must not be overruled by that of the controlling devices;
- levers and other controlling and setting devices must be clearly marked;
- gas leakage rate is not dangerous;
- gas release during ignition and re-ignition and after flame extinction is limited;

- dangerous accumulation of unburned gas is limited;
- ignition and re-ignition is smooth;
- cross-lighting is assured;
- flame stability is assured and combustion products do not contain unacceptable concentrations of substances harmful to health;
- when used normally, there will be no accidental release of combustion products;
- even in abnormal draught conditions there is no release of combustion products;
- independent flueless domestic heating appliances and/or flueless instantaneous water heaters must not cause a carbon monoxide concentration likely to present a danger to the health of persons exposed;
- rational use of energy shall be assured;
- the temperature of the appliance shall not reach temperatures which present a danger in the surrounding area;
- surface temperature of knobs and levers of appliances intended to be manipulated must not present a danger to the user;
- surface temperatures of external parts must not under operating conditions present a danger to the user (in particular to children);
- materials and components which may come into contact with food or water used for sanitary purposes, must not impair their quality.

4.4.7.6 Proof of Conformity

Conformity certification of series-manufactured appliances consists of EC-Type Examination plus (at the choice of the manufacturer):

- the EC Declaration of Conformity; or
- the EC Declaration of Conformity to type (guarantee of production quality); or

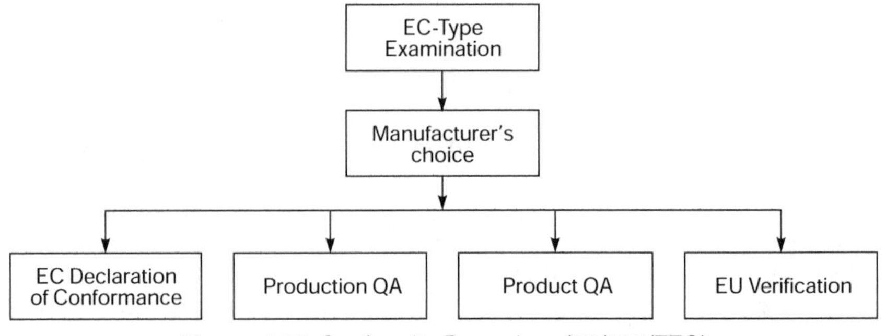

Figure 4.16 Conformity Procedure (90/396/EEC)

Title	*Yes/No*
CE Marking	Yes
EC Declaration of Conformity	Yes
EC Certificate of Conformity	
EC-Type Examination	Yes
Product Verification	Yes
Technical File	Yes
Unit Verification	Yes

Table 4.26 Proof of Conformity (90/396/EEC)

- the EC Declaration of Conformity to type (guarantee of product quality); or
- EC Verification.

Conformity certification in respect of single unit or small quantity production shall be EC verification by unit.

4.4.7.6.1 CE Marking

The appliance or its data plate must bear the CE Marking together with the following inscriptions:

- the manufacturer's name or identification symbol;
- the trade name of the appliance;
- the type of electrical supply used, if applicable;
- the appliance category;
- the last two digits of the year in which the CE Marking was affixed;

Information needed for installation purposes may be added according to the nature of the appliance.

The CE Marking and the inscriptions must be affixed in a visible, easily legible and indelible form to the appliance or to a data plate attached to it. The data plate shall be so designed that it cannot be re-used.

4.4.8 Energy Efficiency Requirements for Household Electric Refrigerators, Freezers and Combinations Thereof (96/57/EC)

Directive:	96/57/EC Dated: 3 Sep 96
Short name	Energy efficiency requirements for household refrigerators and freezers
Amendments	Nil
Repealed Directives	Nil
Member States Implementation Date	23 Sep 96
Transitional Period	Until 18 Oct 02, all Instruments that conformed to the rules in force before the application date
Date of entry into force	23 Sep 96

4.4.8.1 Introduction

This Directive has been issued as part of the programme to promote energy efficiency in the EU, or the SAVE programme, as it is commonly known.

The aims of the programme are to:

- promote the more rational use of energy by the further development of integrated energy-saving policies;
- improve the efficiency of final energy demand by at least 20 per cent;
- ensure a prudent and rational utilisation of natural resources;
- take proper account of potential climatic change linked to the greenhouse effect;
- stabilise (by year 2000) CO_2 emissions at their 1990 level.

4.4.8.2 Structure

The Directive includes the Annexes listed in Table 4.27.

4.4.8.3 Objectives

The prime objectives of the SAVE programme are to:

- encourage consumers to favour appliances and equipment with high electrical efficiency;
- improve the efficiency of appliances and equipment.

Annex	*Title*
I	Method for calculating the maximum allowable electricity consumption of a refrigeration appliance and procedure for checking conformity
II	Conformity Assessment Procedures (Module A)
III	CE Marking

Table 4.27 Annexes to Directive 96/57/EC

4.4.8.4 Exclusions

- Refrigeration equipment for commercial use.
- Appliances which are also capable of using other energy sources – particularly accumulators and household refrigeration appliances working on the absorption principle.
- Appliances manufactured on a one-off basis shall be excluded.

4.4.8.5 Essential Requirements

None stated other than the overriding requirement for energy efficiency.

4.4.8.6 Proof of Conformity

The manufacturer must affix the CE Marking to each refrigeration appliance that he manufactures and draw up a written Declaration of Conformity.

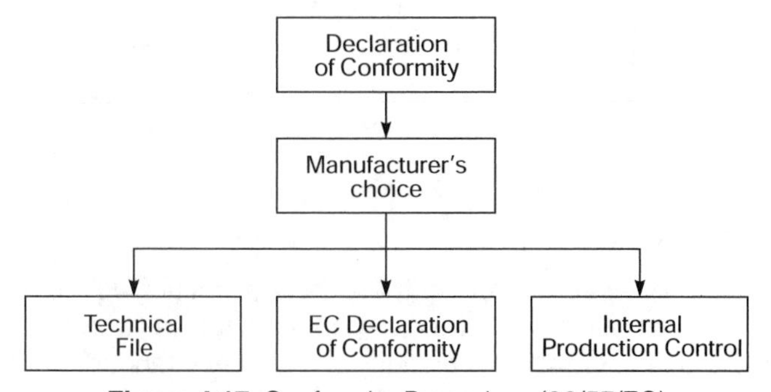

Figure 4.17 Conformity Procedure (96/57/EC)

Title	*Yes/No*
CE Marking	Yes
EC Declaration of Conformity	Yes
Technical File	Yes
Internal Production Control	Yes

Table 4.28 Proof of Conformity (96/57/EC)

4.4.8.7 Other relevant information

Before the expiry of a period of four years from the adoption of this Directive, the Commission shall make an assessment of the results obtained as compared with those expected.

4.4.9 Safety of Toys (89/376/EEC)

Directive:	**89/376/EEC**	**Dated: 3 May 88**
Short name	Toys or Toy Safety	
Description	Safety of toys for children under 14 years of age	
Amendments	93/68/EEC (CE Marking)	
Repealed Directives		
Member States Implementation Date	30 Jun 89 Implemented in the UK by the Toys (Safety) Regulations 1989 (SI 1989:1275)	
Transitional Period		
Date of entry into force	1 Jan 90	

4.4.9.1 Introduction

Toys was the first type of product for which Essential Requirements were adopted. The overall, binding requirement is that toys (which are taken to mean any product, or material designed or clearly intended for use in play by children of less than 14 years of age), may only be placed on the market (both for sale and distribution free of charge) if they do not jeopardise the safety and health of users or third parties when they are used as intended or in a foreseeable way – bearing in mind the normal behaviour of children.

4.4.9.2 Structure

The Directive includes the Annexes listed in Table 4.29.

Annex	*Title*
I	Products not regarded as toys for the purposes of this Directive
II	Essential Safety requirements for toys
III	Conditions to be fulfilled by Approved Bodies
IV	Warnings and indications of precautions to be taken when using toys

Table 4.29 Annexes to Directive 89/376/EEC

4.4.9.3 Objectives

To harmonise the safety regulations for toys throughout the EU in order to protect child health and safety and to facilitate trade.

4.4.9.4 Exclusions

The following are excluded from the scope of the Toy Directive:

- Christmas decorations;
- detailed scale models for adult collectors;
- equipment intended to be used collectively in playgrounds;
- sports equipment;
- aquatic equipment intended to be used in deep water;
- folk dolls and decorative dolls and other similar articles for adult collectors;
- 'professional' toys installed in public places (shopping centres, stations, etc.);
- puzzles with more than 500 pieces or without picture, intended for specialists;
- air guns and air pistols;

- fireworks, including percussion caps (with the exception of percussion caps specifically designed for use in toys without prejudice to more stringent provisions already existing in certain Member States);
- slings and catapults;
- sets of darts with metallic points;
- electric ovens, irons or other functional products operated at a nominal voltage exceeding 24 volts;
- products containing heating elements intended for use under the supervision of an adult in a teaching context;
- vehicles with combustion engines;
- toy steam engines;
- bicycles designed for sport or for travel on the public highway;
- video toys that can be connected to a video screen, operated at a nominal voltage exceeding 24 volts;
- babies' dummies;
- faithful reproductions of real fire arms;
- fashion jewellery for children.

4.4.9.5 Essential Requirements

Toys placed on the market should not jeopardise the safety and/or health, either of users or of third parties.

Note: The standard of safety of toys should be determined in relation to the criterion of the intended use of the product with allowance being made for any foreseeable use (bearing in mind the normal behaviour of children who do not generally show the same degree of care as the average adult user!).

The standard of safety of the toy must be considered when it is marketed, bearing in mind the need to ensure that this standard is maintained throughout the foreseeable and normal period of use of the toy.

The Essential Requirements are divided into 'General Principles' and 'Particular Risks'.

4.4.9.5.1 General Principles

The users of toys as well as third parties must be protected against health hazards and risk of physical injury when toys are used as intended or in a foreseeable way, bearing in mind the normal behaviour of children. Such risks are those which are:

- connected with the design, construction or composition of the toy;
- inherent in the use of the toy.

The degree of risk present in the use of a toy must be commensurate with the ability of the users, and where appropriate their supervisors, to cope with it.

A minimum age for users of toys and/or the need to ensure that they are used only under adult supervision must be specified where appropriate.

Labels on toys and/or their packaging and the instructions for use which accompany them must draw the attention of users or their supervisors fully and effectively to the risks involved in using them and to the ways of avoiding such risks.

4.4.9.5.2 Particular Risks

The Essential Requirements shown in the Directive cover:

- physical and mechanical properties;
- flammability;
- chemical properties;
- electrical properties;
- hygiene;
- radioactivity.

4.4.9.6 Proof of Conformity

In order to affix the CE Marking, the manufacturer is required to produce a Technical File and obtain an EC-Type Examination certificate.

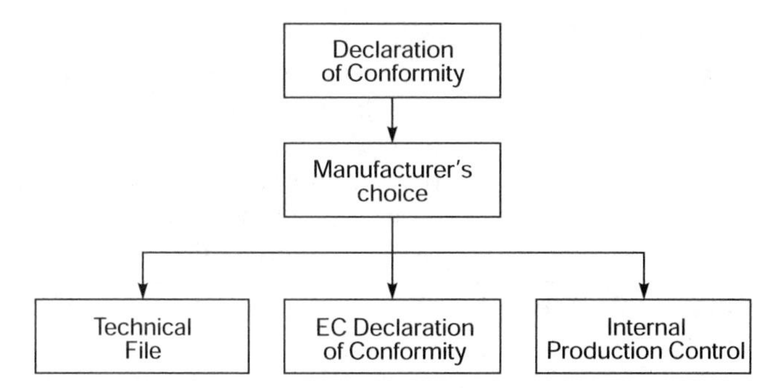

Figure 4.18 Conformity Procedure (89/376/EEC)

Title	*Yes/No*
CE Marking	Yes
EC Declaration of Conformity	Yes
EC-Type Examination	Yes
Technical File	Yes

Table 4.30 Proof of Conformity (89/376/EEC)

4.4.9.6.1 CE Marking

CE Marking on a toy indicates that **all** the Essential Requirements of **all** of the applicable Directives have been met.

4.4.10 Recreational Craft (94/25/EEC)

Directive:	**94/25/EEC** **Dated: 19 Feb 73**
Short name	Recreational Craft Directive
Description	Design and construction of boats of 2.5 to 24 m, plus specified components, excluding hydrofoils and hover craft, and craft for charter
Amendments	93/68/EEC (CE Marking)
Repealed Directives	Nil
Member States Implementation Date	16 Dec 95 Implemented in the UK by the Recreational Craft Regulations 1996 (SI 1996:1353)
Transitional Period	A transitional period ending on 30 June 1998 is fixed by the directive during which the placing on the market and/or putting into service of products in conformity with national legislation in force before 30 June 1994 is authorised
Date of entry into force	15 Jun 94

4.4.10.1 Introduction

In view of the possible risks involved by using recreational craft, procedures for ensuring compliance with the Directive's Essential Requirements have to be applied. These procedures need to be related to the level of risk that may be inherent in the recreational craft and their associated components.

This Directive applies to all recreational craft, partly completed boats (and components), when separate and when installed.

4.4.10.2 Structure

The Directive includes the Annexes listed in Table 4.31.

Annex	Title
I	Essential Safety Requirements for the design and construction of Recreational Craft
II	Components
III	Declaration by the builder or his authorised representative established in the Community or the person responsible for placing on the market
IV	CE Marking
V	Internal Production Control (Module A)
VI	Internal Production Control plus tests (Module AA option 1)
VII	EC-Type Examination
VIII	Conformity to Type
IX	Production Quality Assurance

Table 4.31 Annexes to Directive 94/25/EEC

4.4.10.3 Objectives

To harmonise the essential health and safety requirements for recreational boats (i.e. any boat from 2.5 m to 24 m hull length intended for sports and leisure purposes) and components (when separate and when installed) throughout the Community in order to protect the health and safety of persons, property or the environment and to facilitate trade.

Note: 'Components' in the context of this Directive means:

- ignition-protected equipment for inboard and stern drive engines;
- start-in-gear protection devices for outboard engines;
- steering wheels, steering mechanisms and cable assemblies;
- fuel tanks and fuel hoses;
- prefabricated hatches and portlights.

4.4.10.4 Exclusions

Craft excluded from the scope of the directive include:

- craft intended solely for racing (including racing and training rowing boats, which are labelled as such by the manufacturer);
- canoes and kayaks;
- gondolas and pedalos;
- sailing surfboards;

- personal watercraft (including powered surfboards);
- original and individual replicas of historic craft designed before 1950;
- experimental craft that are not placed on the EU market;
- craft built for own use (provided that they are not placed on the market within a period of five years);
- craft intended to carry passengers for commercial purposes;
- submersibles;
- air cushion vehicles; and
- hydrofoils.

4.4.10.5 Essential Requirements

Boats in each Category (i.e. Ocean, Offshore, Inshore or Sheltered Waters) must be designed and constructed to meet the relevant Essential Requirements listed in the Directive and to have good handling characteristics.

This Directive does not contain any provisions that directly limit the use of the recreational craft after it has been put into service.

4.4.10.5.1 General Requirements

- Hull identification – each craft must be marked with a hull identification number that includes the:
  - manufacturer's code;
  - country of manufacture;
  - unique serial number;
  - year of production;
  - model year.
- Builder's plate – each craft is required to carry a permanently affixed plate mounted separately from the boat hull identification number, containing the following information:
  - manufacturer's name;
  - CE Marking;
  - boat design category;
  - manufacturer's maximum recommended load;
  - number of persons the craft is recommended to carry.
- Protection from falling overboard and means of reboarding;
- Visibility from the main steering position;
- Owners manual.

4.4.10.5.2 Integrity and Structural Requirements

The following requirements are detailed in the Directive:

- structure;
- stability and freeboard;
- buoyancy and flotation;

- openings in hull, deck and superstructure;
- flooding – all craft shall be designed so as to minimise the risk of sinking;
- manufacturer's maximum recommended load;
- life raft stowage;
- escape – all habitable craft shall be provided with viable means of escape in the event of fire, or inversion;
- anchoring, mooring and towing.

4.4.10.5.3 Handling Characteristics

The manufacturer is required to ensure that the handling characteristics of the craft are satisfactory with the most powerful engine for which the boat is designed and constructed.

4.4.10.5.4 Installation Requirements

The following installation requirements are stipulated:

- engines and engine spaces;
- inboard engine;
- ventilation;
- exposed parts;
- outboard engines starting;
- fuel system and tanks;
- electrical system;
- steering system;
- emergency arrangements;
- gas system;
- fire protection and fire-fighting equipment;
- navigation lights;
- discharge prevention.

4.4.10.6 Proof of Conformity

The requirements for Proof of Conformity in this Directive are split into four categories (i.e. craft designed for Ocean, Offshore, Inshore or Sheltered Waters). The following is the basic requirement:

- Ocean and Offshore craft:
  - boats less than 12 m hull length – Internal Production Control;
  - boats from 12 m to 24 m hull length: – EC-Type Examination supplemented by Conformity to Type.
- Inshore craft:
  boats less than 12 m hull length – Internal Production Control (provided that the craft complies with the harmonised standards) or

Internal Production Control plus some additional tests if it does not comply;
  - boats from 12 m to 24 m hull length EC-Type Examination followed by Conformity to Type.
- Sheltered Water craft:
  - boats from 2.5 m to 24 m hull length – Internal Production Control.

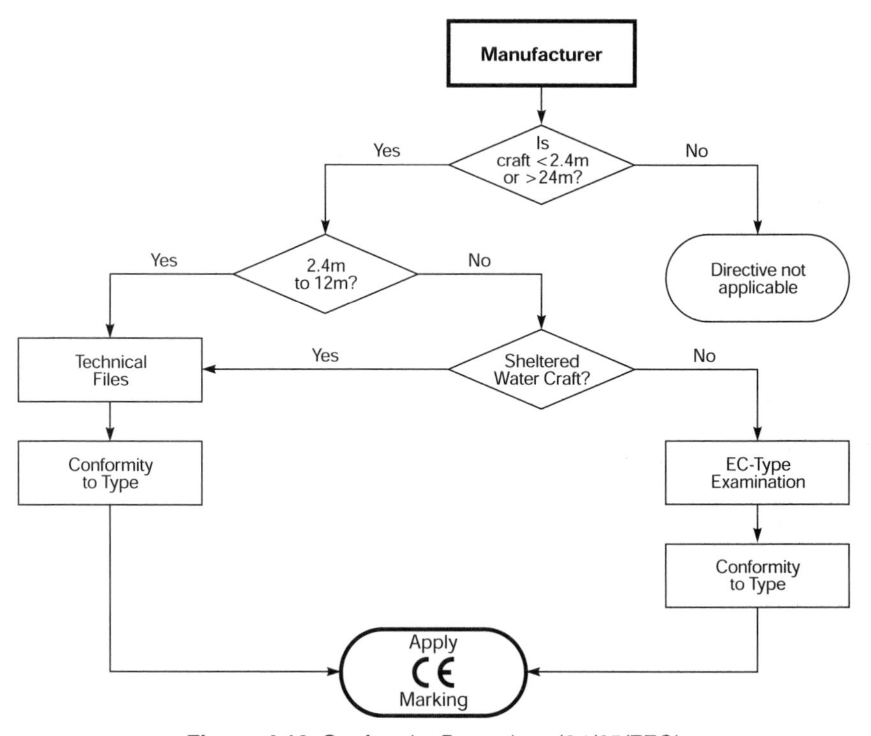

Figure 4.19 Conformity Procedure (94/25/EEC)

Title	*Yes/No*
CE Marking	Yes
EC Declaration of Conformity	Yes
EC-Type Examination	Yes
Production Quality Assurance	Yes
Technical File	Yes
Internal Production Control	Yes
Conformity to Type	Yes

Table 4.32 Proof of Conformity (94/25/EEC)

4.4.10.6.1 Publications

Application Guide '*Recreational Craft Directive and comments to the Directive combined*'.

4.4.10.6.2 CE Marking

The CE Marking and the Builder's Plate, including the name and address of the manufacturer (or his authorised representative in the EU), shall be on the recreational craft in easily legible and indelible form and the CE Marking shall be on components as referred to in Annex II or on their packaging.

4.4.11 New Hot-Water Boilers Fired with Liquid or Gaseous Fluids (Efficiency Requirements) (92/42/EEC)

Directive:	**92/42/EEC** **Dated: 92/42/EEC**
Short name	Hot-Water Boilers
Description	Efficiency requirements for hot-water boilers of 4 to 400 kW output
Amendments	93/68/EEC (CE Marking)
Repealed Directives	Nil
Member States Implementation Date	1 Jan 93 Implemented in the UK by the Boiler (Efficiency) Regulations 1993 (SI 1993:3083)
Transitional Period	Until 31 Dec 97, all Instruments that conformed to the rules in force before the application date
Date of entry into force	1 Jan 94

4.4.11.1 Introduction

This Directive has been issued as part of the programme to promote energy efficiency in the EU or the SAVE programme, as it is commonly known. The aims of the programme are to:

- promote the more rational use of energy by the further development of integrated energy-saving policies;
- improve the efficiency of final energy demand by at least 20 per cent;
- ensure a prudent and rational utilisation of natural resources;

- take proper account of potential climatic change linked to the greenhouse effect;
- stabilise (by year 2000) CO_2 emissions at their 1990 level.

4.4.11.2 Structure

The Directive includes the Annexes listed in Table 4.33.

Annex	*Title*
I	Conformity Marks and additional specific Markings
II	Award of energy-performance labels
III	EC-Type Examination
IV	Conformity to Type Production Quality Assurance Product Quality Conformance
V	Minimum Criteria to be taken into account by Member States for the Notification of Bodies

Table 4.33 Annexes to Directive 92/42/EEC

4.4.11.3 Objectives

To determine the requirements for better efficiency of boilers in the interest of consumers and energy preservation and to eliminate technical barriers with regard to boiler efficiency.

This Directive is part of SAVE programme concerning the promotion of energy efficiency in the EU and details the efficiency requirements that are applicable to new hot-water boilers fired by liquid or gaseous fuels with a rated output of no less than 4 kW and no more than 400 kW, hereinafter called 'boilers'.

4.4.11.4 Exclusions

The following are excluded from this Directive:

- hot-water boilers capable of being fired by different fuels including solid fuels;
- equipment for the instantaneous preparation of hot water;
- boilers designed to be fired by fuels, the properties of which differ appreciably from the properties of the liquid and gaseous fuels commonly marketed (industrial waste gas, biogas, etc.);
- cookers and appliances designed mainly to heat the premises in which they are installed and, as a subsidiary function, to supply hot water for central heating and sanitary hot water;

- appliances with rated outputs of less than 6 kW using gravity circulation and designed solely for the production of stored sanitary hot water;
- boilers manufactured on a one-off basis.

4.4.11.5 Essential Requirements

The various boiler types must comply with the useful efficiency requirements for rated output (i.e. Pn expressed in kW at an average boiler-water temperature of 70°C) and at part load (i.e. operating at 30 per cent).

The efficiency of the boiler is shown by a series of stars as follows:

- if its efficiency at rated output and at part load are equal to or greater than the relevant values for standard boilers, a boiler is awarded one star;
- if its efficiency at rated output and at part load are three or more points higher than the relevant values for standard boilers, a boiler is awarded two stars;
- every extra step of efficiency of three points at rated output and at part load is granted an additional star.

4.4.11.6 Proof of Conformity

The conformity of series-produced boilers is certified by EC-Type Examination followed by an EC Declaration of Conformity according to which route it follows, i.e.:

- Module C – Conformity to Type;
- Module D – Production Quality Assurance;
- Module E – Product Quality Assurance.

4.4.11.6.1 CE Marking

CE Marking is accompanied by a star symbol (see Figure 4.21) indicating its energy efficiency.

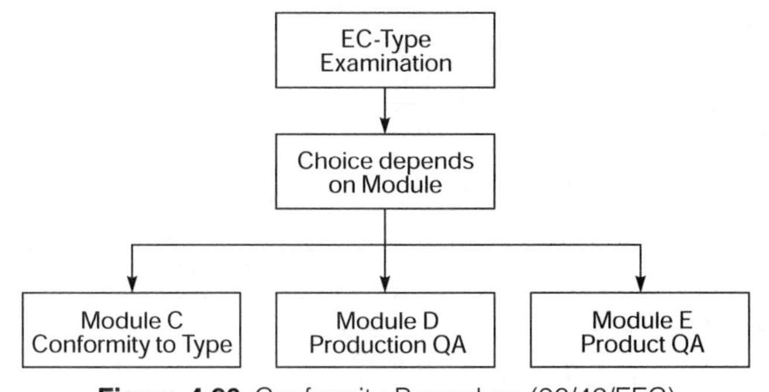

Figure 4.20 Conformity Procedure (92/42/EEC)

Title	*Yes/No*
CE Marking	Yes
EC Declaration of Conformity	Yes
EC Certificate of Conformity	Yes
EC-Type Examination	Yes
Product Quality Assurance	Yes
Production Quality Assurance	Yes
Technical File	Yes

Table 4.34 Proof of Conformity (92/42/EEC)

Figure 4.21 Boiler energy-efficiency symbol

4.4.12 Construction Products (89/106/EEC)

Directive:	**89/106/EEC** **Dated: 21 Dec 88**
Short name	Construction Products
Description	Safety and performance of building products including requirements for mechanical stability, fire resistance, hygiene, noise and energy efficiency
Amendments	93/68/EEC (CE Marking)
Repealed Directives	Nil
Member States Implementation Date	27 Jun 91 Implemented in the UK by the Construction Products Regulations 1991 (SI 1991:1620)
Transitional Period	Nil
Date of entry into force	27 Jun 91

4.4.12.1 Introduction

This Directive applies to all construction products that are used as part of building and civil engineering works.

4.4.12.2 Structure

The Directive includes the Annexes listed in Table 4.35.

Annex	*Title*
I	Essential Requirements
II	European Technical Approval
III	Attestation of Conformity with the Technical Specifications
IV	Approval of Testing Laboratories, Inspection Bodies and Certification Bodies

Table 4.35 Annexes to Directive 89/106/EEC

4.4.12.3 Objectives

To harmonise national legislation on products with respect to the health and safety requirements applicable to construction.

4.4.12.4 Essential Requirements

The Essential Requirements applicable to works which may influence the technical characteristics of a product are set out in terms of objectives in Annex I of the Directive. These concern:

- mechanical resistance and stability;
- safety in case of fire;
- hygiene, health and the environment;
- safety in use;
- protection against noise;
- energy economy and heat retention.

One, some or all of these requirements may apply; they shall be satisfied during an economically reasonable working life.

4.4.12.5 Proof of Conformity

The manufacturer is responsible for the attestation that products are in conformity with the requirements of the appropriate technical specification

Title	*Yes/No*
CE Marking	Yes
EC Declaration of Conformity	Yes
Technical File	Yes

Table 4.36 Proof of Conformity (89/106/EEC)

and all products that are the subject of an attestation of conformity shall benefit from the presumption of conformity.

The attestation of conformity of a product is dependent on:

- the manufacturer having a factory production control system to ensure that production conforms with the relevant technical specifications; or
- whether it is a particular product (as indicated in the relevant technical specifications), in which case, in addition to a factory production control system, an approved certification body is involved in the assessment and surveillance of the production control or of the product itself;
- in the case of individual (and non-series) production, a Declaration of Conformity in accordance with Annex III of the Directive (i.e. the Attestation of Conformity with the Technical Specifications) will suffice.

All that is required to enable the manufacturer to affix the CE Marking is to make a Declaration of Conformity or issue a Certificate of Conformity.

4.4.12.5.1 Publications

Application Guide '*Guidance papers concerning Construction Products Directive 89/106/EC*'.

4.4.12.5.2 CE Marking

The following additional information regarding the affixing of CE Marking was made by the 'CE Amending Directive' (93/68/EEC):

- the CE Marking shall be followed by the identification number of the body involved in the production control stage;
- the CE Marking shall be accompanied by the name or identifying mark of the producer, the last two digits of the year in which the Marking was affixed, and where appropriate, the number of the EC Certificate of Conformity and, where appropriate, indications to identify the characteristics of the product on the basis of the technical specifications.

4.4.13 Lifts (95/16/EEC)

Directive:	**95/16/EEC** **Dated: 29 Jun 95**
Short name	Lifts
Description	Safety of lifts for carrying people
Amendments	
Repealed Directives	84/528/EEC (Common provisions for lifting and mechanical handling appliances) 84/529/EEC (Electrically, hydraulically or oil-electrically operated lifts)
Member States Implementation Date	1 Jan 97 No additional UK legislation required for implementation although modification to some existing legislation is being considered
Transitional Period	Until 30 Jun 99, all lifts and safety components that conformed to the rules in force before the application date
Date of entry into force	1 Jul 97

4.4.13.1 Introduction

Directive 95/16/EC updates and improves the existing Directive relating to 'common provisions for lifting and mechanical handling appliances' (i.e. 84/528/EEC) which serves as a framework Directive for two specific lift Directives, namely Directive 84/529/EEC (relating to electrically, hydraulically or oil-electrically operated lifts) and Directive 86/663/EEC (relating to self-propelled industrial trucks).

4.4.13.2 Structure

The Directive includes the Annexes listed in Table 4.37.

4.4.13.3 Objectives

To preserve the health and safety of the users of **permanently** installed lifts and of people occupying the environment of lifts.

Annex	*Title*
I	Essential health and safety requirements relating to the design and construction of lifts and safety components
II	EC Declaration of Conformity
II–A	EC Declaration of Conformity – for Safety Components
II–B	EC Declaration of Conformity – for installed lifts
III	CE Conformity Marking
IV	List of Safety Components Referred to in Article 1(1) and 8(1) of the Directive
V	EC-Type Examination
V–A	EC-Type Examination – Safety Components
V–B	EC-Type Examination – Lifts
VI	Final Inspection
VII	Minimum criteria to be taken into account by Member States for the Notification of Bodies
VIII	Product Quality Assurance
IX	Full Quality Assurance
X	Unit Verification
XI	Conformity to type with random checking
XII	Product Quality Assurance for Lifts
XIII	Full Quality Assurance for Lifts
XIV	Production Quality Assurance

Table 4.37 Annexes to Directive 95/16/EEC

To avoid obstruction to and impeding access and use by disabled persons and to allow any adjustment facilitating the use of lifts by disabled people.

To cover defined safety components of lifts.

Note: Safety components are listed in the Directives as:

- devices for locking doors;
- devices to prevent falls;
- energy accumulating shock absorbers;
- jacks of hydraulic power circuits;
- safety switches.

4.4.13.4 Exclusions

This Directive does not apply to:

- cableways, including funicular railways, for the public or private transportation of persons;
- lifts specially designed and constructed for military or police purposes;
- mine winding gear;
- theatre elevators;
- lifts fitted in means of transport;
- lifts connected to machinery and intended exclusively for access to the workplace;
- rack and pinion trains;
- construction-site hoists intended for lifting persons or persons and goods.

4.4.13.5 Essential Requirements

The Essential Requirements are listed under six main headings, namely:

- **General** – car, suspension and support, loading control, machinery, controls, electrical equipment;
- **Hazards to persons outside the car** – risk of crushing, landings, interlocking devices;
- **Hazards to persons in the car** – enclosures, power cuts/failure, buffers;
- **Other Hazards** – landing doors, counterweights, temperature of machinery, ventilation for passengers, emergency lighting, communication, control circuits;
- **Marking** – each car must bear an easily visible plate clearly showing the rated load, maximum number of passengers, emergency escape instructions;
- **Instructions for use** – a manual detailing assembly, connection, adjustment and maintenance.

Note

1. **Safety Comments** – must be accompanied by an instruction manual to enable assembly, connection, adjustment and maintenance to be carried out effectively.
2. **Lifts** – must be accompanied by documentation containing (as a minimum):
   - an instruction manual containing the plans and diagrams necessary for normal use and relating to maintenance, inspection, repair, periodic checks and rescue operations;
   - a logbook in which repairs and, where appropriate, periodic checks can be noted.

3 **89/392/EEC** – where the relevant hazard exists and is not dealt with in the Essential Requirements of this Directive, the essential health and safety requirements of Annex I to Directive 89/392/EEC are applicable.

4.4.13.6 Proof of Conformity

4.4.13.6.1 Safety Components

The manufacturer must draw up a Declaration of Conformity and either:

- submit the model of the safety component for EC-Type Examination; or
- submit the model of the safety component for EC-Type Examination **and** operate a Product Quality Assurance system; or
- operate a Full Quality Assurance system.

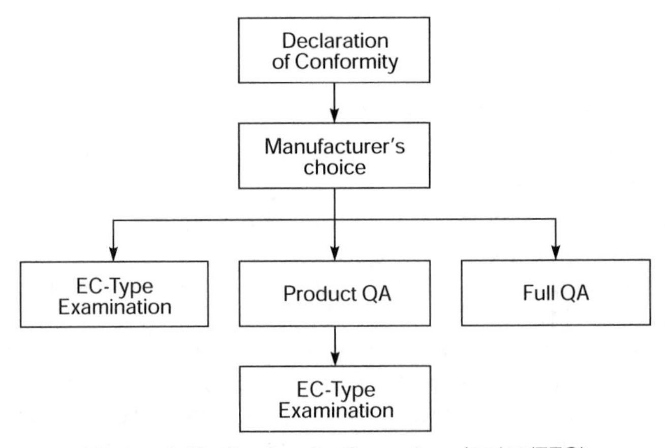

Figure 4.22 Conformity Procedure (95/16/EEC)

Title	*Yes/No*
CE Marking	Yes
EC Declaration of Conformity	Yes
EC-Type Examination	Yes
Product Quality Assurance	Yes
Production Quality Assurance	Yes
Full Quality Assurance	Yes
Technical File	Yes
Unit Verification	Yes
Conformity to Type	Yes

Table 4.38 Proof of Conformity (95/16/EEC)

4.4.13.6.2 Lifts

The manufacturer must submit the model for EC-Type Examination and then either:

- subject the lift to a final inspection (as detailed in Annex VI of the Directive); or
- implement a Product Quality Assurance system; or
- a Full Quality Assurance system; or
- if it was designed in accordance with a model lift having already undergone an EC-Type Examination, it shall be constructed, installed and tested by implementing:
  - the final inspection (as detailed in Annex VI of the Directive); or
  - operate a Product Quality Assurance system; or
  - operate a Production Quality Assurance system; or
- if it was designed in accordance with a lift for which a quality assurance system was implemented, it shall be installed and constructed and tested by implementing:
  - the final inspection (as detailed in Annex VI of the Directive); or
  - the Product Quality Assurance system; or
  - the Production Quality Assurance system; or
- having undergone the Unit Verification procedure, by a Notified Body; or
- having been subject to a Full Quality Assurance system supplemented by an examination of the design if the latter is not wholly in accordance with the harmonised standards.

4.4.13.7 Other relevant information

No later than 30 June 2002, the Commission shall re-examine the functioning of the procedures laid down in this Directive and, if necessary, submit proposals for appropriate amendments.

4.5 Other Directives associated with the CE Marking Directive

4.5.1 Product Liability Directive (85/374/EEC)

The Product Liability Directive is applicable to all products covered by New Approach Directives. Its objective is to:

- protect public interest (e.g. health and safety of persons, consumer protection, protection of business transactions, environmental protection);
- prevent, as far as possible, the placing on the market and putting into service of unsafe or otherwise non-compliant products;

- force the supplier to make safe products in order to avoid the costs that liability places on him for defective products.

The Product Liability Directive requires that all producers involved in the production process should be made liable, for their finished product, component part, or any raw material supplied to them which was defective. The producer shall not be liable if he proves that the defect is due to compliance of the product with mandatory regulations issued by the public authorities.

To protect the physical well-being and property of the consumer, the defectiveness of the product should be determined by reference to its fitness for use and to the degree of safety which the public at large is entitled to expect (e.g. harmonised or national standards).

4.5.2 General Product Safety Directive (92/59/EEC)

The General Product Safety Directive (92/59/EEC) requires product manufacturers and suppliers to 'place only "safe products" on the market' and to ensure a 'high level of protection of safety and health of persons'.

All consumer products not complying with the definition laid down by the Directive in question, used and second-hand products that were originally placed on the EU market before the directive entered into force, or repaired products, will come under the general product safety Directive.

When there are no specific EU provisions for safety of a product (i.e., EU laws and standards), the conformity of a product to the general safety requirement shall be assessed against voluntary national standards, or, failing these, to standards drawn up in the Member State in which the product is in circulation.

4.6 Model EC Declaration of Conformity

The manufacturer or his authorised representative established in the Community ([1]):

____

____

____

declares that the industrial product described hereafter ([2])

____

____

____

is in conformity with the provisions of Council Directive ______ and, where such is the case, with the national standard transposing harmonised standard No ______ is identical to the industrial product which is the subject of EC Certificate of Conformity No ______ issued by ([3]) ([4]) ________________________

____

____

is subject to the procedure set out in Article ______ of Directive ______ under the supervision of the Notified Body ([3]) __________

Done ____________________, on ____________________

____

Signature([3])

(1) Business name and full address, authorised representatives must also give the business name and address of the manufacturer.
(2) Description of the industrial product (make, type, serial number, etc.)
(3) Name and address of the approved body.
(4) Delete whichever is inapplicable.
(5) Name and position of the person empowered to sign on behalf of the manufacturer or his authorised representative.

5 GAINING CE CONFORMITY

As shown in Figure 5.1, there are three methods of gaining CE Conformance.

For most products and machines, the self-declaration process (module A) is possible. In practice, the manufacturer performs the complete product assessment according to EU standards, issues the declaration, and affixes the CE Marking to the product. A Technical File or documentation must also be available on demand for national enforcement authorities.

5.1 Self-declaration

Under self-declaration, the manufacturer takes complete responsibility for the assessment, testing, documentation, and declaration of conformity and CE Marking. This route is available for products/machines where a mandatory type exam is not required.

5.2 Voluntary Certification

Frequently, manufacturers have products assessed by a European Body (Notified or Competent) for a 'mark and certification' for marketing purposes and as a defence of 'due diligence' in the event of a challenge.

5.3 Mandatory Certification

Most products and machines do not require mandatory certification. But in such cases, as with some high-risk machinery or when harmonised

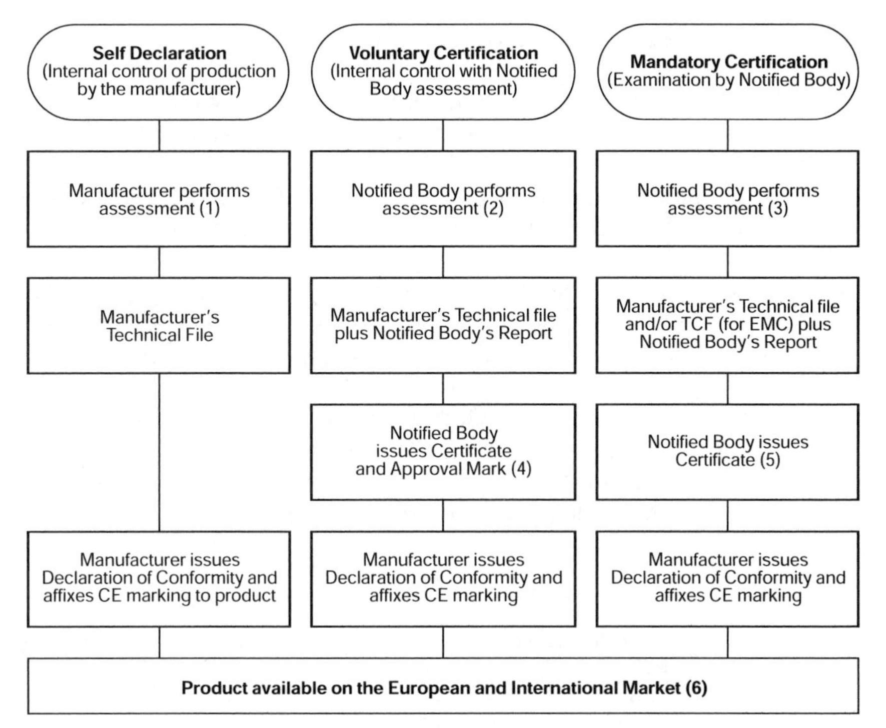

Figure 5.1 Conformity assessment procedures

Notes:

1 Manufacturer's assessment of the Directive on the basis of conformity with European harmonised standards and/or the Essential Requirements.
2 Notified Body's assessment based on European harmonised standards (or other criteria when no harmonised or equivalent national standard is available).
3 Notified Body's assessment and approval of the manufacturer's Technical File based on European harmonised standards (or other criteria when no harmonised or equivalent national standard is available).
4 Subject to random product check by the Notified Body.
5 Approval Mark available. Subject to random product check by the Notified Body.
6 A Technical File (and/or TCF in the case of an EMC product), plus a Declaration of Conformity and CE Marking is required before shipment.

standards do not exist or are not applied in full, a 'type examination' by a Notified Body is required. After successful testing, a Type-Examination Certificate (for machinery) or Certificate of Conformity (for EMC) is issued by the EU Body. The manufacturer then affixes the CE Marking and issues a declaration of conformity.

The Declaration of Conformity procedures are listed in the relevant Directives.

5.4 The Five Steps to Conformity

Regardless of whether the manufacturer is following the self-declaration route or using an Accredited Body for voluntary or mandatory certification procedures, all five steps are required.

Thus to achieve its main objectives, the CE Directive requires manufacturers of all products covered by the Directive to:

- follow an **EC Declaration of Conformity**. This is the procedure whereby the manufacturer or authorised representative 'ensures and declares' that the products concerned satisfy the requirements of the Directives that apply to them;
- possess a fully auditable **Quality Management System** (QMS) consisting of Quality Policies, Quality Procedures and Work Instructions. This is the means by which a manufacturer can prove he is conforming to the requirements of the CE Directive.

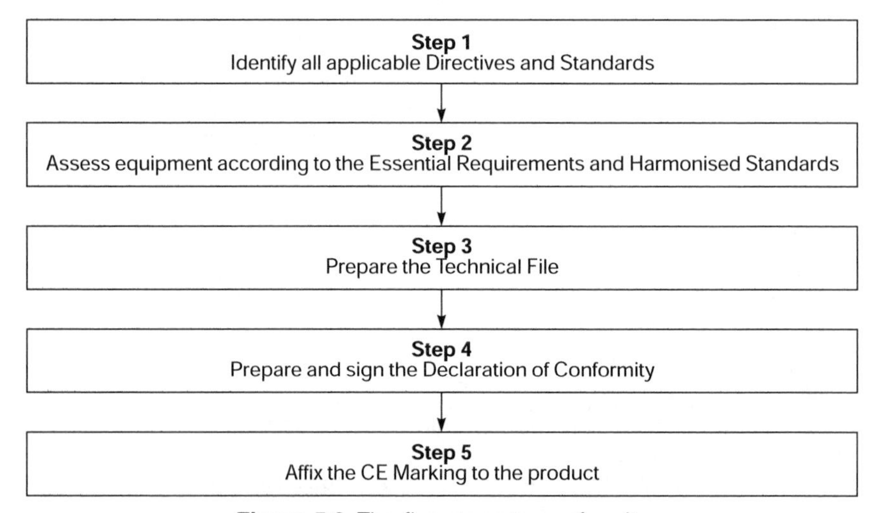

Figure 5.2 The five steps to conformity

5.5 EC Declaration of Conformity

Before an industrial product may be offered for sale and/or use, the manufacturer must draw up a written Declaration of Conformity. This Declaration of Conformity is the EC's procedure whereby the manufacturer ensures and declares that the industrial product they have manufactured meets the requirements and provisions of the Directive, which apply to this product.

The Declaration of Conformity shall contain:

- the name and address of manufacturer or representative;
- description of the product (name, model number, etc.);
- Directives(s) declared;
- harmonised standards applied;
- additional standards and specifications (where appropriate);
- place and date of issue;
- name and signature of authorised person.

The declaration needs to be kept on file, but manufacturers may also supply a declaration with each product or shipment. For machinery, a declaration must accompany each machine. Depending on the actual industrial product, the EC Declaration of Conformity consists of the steps listed in Table 5.1.

Step 1	Documenting the manufacturer's Quality Management System
Step 2	Preparing the manufacturer's technical documentation
Step 3	Application for assessment of a manufacturer's Quality System
Step 4	Audit of the manufacturer's Quality Management System by a Notified Body
Step 5	Application for the examination of a product's design
Step 6	Continued surveillance by the Notified Body
Step 7	Additional (i.e. post-production) requirements

Table 5.1 EC Declaration of Conformity

5.6 Quality Management System

The procedural requirements for a Quality Management System for compliance with the CE Directive have certain things in common, namely:

- documenting the manufacturer's Quality Management System;
- preparing the manufacturer's technical documentation;
- application for assessment of the manufacturer's Quality Management System;
- audit of the manufacturer's Quality Management System by the Notified Body;

- continued surveillance by the Notified Body;
- additional (post production) requirements.

Each of these procedural requirements is explained as follows:

5.6.1 Documenting the Manufacturer's QMS

To meet the requirements of the Directive, the manufacturer's Quality Management System must be documented in the form of written policies and procedures such as quality programmes, plans, manuals and records and, depending on the demands of the Annex, may include:

- the manufacturer's quality objectives;
- the organisational structure, responsibilities, tasks and authority;
- methods for monitoring the Quality Management System (including control of products which fail to conform);
- procedures for monitoring and verifying the design of the products. For example:
  - a general description of the product;
  - design specifications, standards being referred to, risk analysis;
  - techniques used to control and verify the design;
  - if the product is to be connected to another product(s), proof that it conforms to the essential requirements of the other product(s);
  - a statement indicating whether or not the product incorporates, as an integral part, a substance that could be considered as an industrial product in its own right;
  - instructions for use.
- the inspection and quality assurance techniques used during all stages of manufacture and in particular:
  - the processes, procedures (and relevant documents) used for sterilisation;
  - product identification procedures (obtained from drawings, specifications and other relevant documents).
- the appropriate tests and trials which will be carried out before, during and after manufacture, together with their frequency, the equipment used, the requirements for calibrating the test equipment.

5.6.2 Technical Documentation

The manufacturer (or his authorised representative) is responsible for preparing the technical documentation in support of his product. This documentation shall be used for conformance assessment (of the product) by a Notified Body and, depending upon the Annex, may include:

- a general description of the industrial product (including details of any variants);

- design drawings and diagrams of components, sub-assemblies and circuits etc;
- full descriptions and instructions required to fully understand the design drawings/diagrams;
- risk analysis results;
- a list of all the standards used for risk analysis;
- a description of any sterilisation process (if used);
- results of design calculations and any inspections that have been completed;
- if the product is connected to another product, proof that it conforms to the essential requirements and specifications of the second product;
- appropriate test reports and clinical data;
- details of labelling and instructions for use.

5.6.3 Quality Management System – Application Requirements

The manufacturer must apply for an assessment of their Quality Management System by a Notified Body. The application shall include:

- name and address of the manufacturer;
- product or product category;
- a written declaration that no application has been lodged with any other Notified Body for the same product-related quality system;
- a documented Quality Management System;
- an undertaking to fulfil the obligations imposed by the Quality Management System;
- an undertaking to keep the approved quality system adequate and efficacious (i.e. producing the intended results);
- an undertaking to set up and keep a systematic review procedure.

5.6.4 ISO 9000

The Quality System standards that are used with the New Approach Directives for conformity assessment are described in ISO 9001, 9002 and 9003. (Note. These will have to be revised in 2001 to take into consideration the requirements of ISO 9001:2000.)

Compliance with these standards gives a presumption of conformity with the corresponding quality assurance modules (i.e. modules D, E, H and their variants) as regards the provisions covered by the standard in question. However, compliance with modules D, E, H and their variants does not require quality system certification according to standards ISO 9001, 9002 or 9003, although it provides a useful means in establishing compliance.

For the purpose of complying with the applicable directives the manufacturer must ensure that the Quality Management System demonstrates the full application of the essential requirements in question.

5.6.5 Audit of the Manufacturer's Quality Management System by a Notified Body

Having established his Quality Management System (QMS), the next step is for the manufacturer to apply for a Notified Body to complete an audit of his facilities. The aim of this audit is to determine whether the manufacturer's QMS meets the requirements referred to in the relevant harmonised standards.

As well as an inspection of the manufacturer's premises, this audit can also include inspections of the manufacturer's suppliers and/or subcontractors premises and facilities.

The manufacturer **must** inform the Notified Body of any plan for substantial changes to the quality system or the product range covered.

The manufacturer must also lodge (with the Notified Body) an application for examination of the design dossier relating to the product which he plans to manufacture.

This application will describe the design, manufacture and product performances. It will also include the documents required to assess whether the product conforms to the requirements of this Directive.

If the Notified Body considers that the product conforms to the relevant provisions of the CE Directive, he will then issue the applicant with an EC-Design Examination certificate.

Note: any changes to an approved design must be assessed and re-approved by the Notified Body which issued the original EC-Design Examination certificate.

5.6.6 Continued Surveillance by the Notified Body

The aim of pre-planned and agreed surveillance visits to the manufacturer's premises (which are carried out by the Notified Body) is to ensure that the manufacturer continues to fulfil the requirements of his own documented Quality Management System.

Prior to the surveillance, the manufacturer has to supply the Notified Body with all the necessary information. In particular this will include:

- details of his documented Quality Management System;
- data relating to the design of the industrial product, results of analyses, calculation of tests etc.;
- data relating to the actual manufacture – such as inspection reports, test data, calibration data, qualification reports of the personnel concerned, etc.

To make sure that the manufacturer continues to apply his documented (and approved) Quality Management System, the Notified Body will periodically carry out **unannounced** surveillance visits to the manufacturers (and if deemed appropriate) their subcontractors' or suppliers' premises!

5.6.7 Other Considerations

The manufacturer shall:

- complete post-production reviews aimed at identifying any malfunction, deterioration of the product's characteristics or performance as well as any inadequacy concerning the labelling or work instructions issued with the product;
- for a period ending at least five years after the last product has been manufactured, keep at the disposal of the national authorities the following documentation:
  - the declaration of conformity;
  - his approved Quality Management System;
  - details of any changes that have been made to the product following approval;
  - details of the product's design, manufacture and performance;
  - copies of all decisions and reports made by Notified Body, and, for Medical Devices:
- if the product has been placed on the market in a sterile condition the manufacturer is responsible for ensuring the maintenance of those sterile conditions.

5.7 Conformity Assessment Procedures

Table 5.2 summarises the key points and procedures found within each Module.

5.8 Conformance Requirements for Each Module

See Table 5.4.

5.9 Conformance Requirements for New Approach Directives

Table 5.5 is provided as a quick reference to the various Conformance Requirements for New Approach Directives.

Module	*Title*	*Key points*
Module A	Internal Production Control	The manufacturer must draw up a statement for each product giving clear details on traceability and conformance to CE requirements. The manufacturer must produce technical documentation, which would allow assessment of the product to CE requirements.
Module B	EC-Type Examination	The Notified Body ascertains and certifies that a representative sample of the product conforms to the supplied product documentation and fulfils the relevant provisions of the Directive. Manufacturer produces documentation supporting the design, manufacture and performance of a product.
Module D	Production Quality Assurance (IS0 9002:1994)	The manufacturer applies an approved QMS that addresses quality during the manufacture, and final inspection of a product. Involvement of the Notified Body includes approval of the QMS for manufacture through to completion and post-production follow up.
Module E	Product Quality Assurance (ISO 9003:1994)	The manufacturer applies an approved QMS that addresses quality for the final inspection and testing of a product. Involvement of the Notified Body includes approval of the QMS for final inspection and testing, through to post-production follow up.
Module F	Product Verification	The manufacturer, or his authorised representative, ensures and declares that the industrial products that he is producing conform to the type described in the EC-type examination certificate and meet the applicable CE requirements.

Table 5.2 Key points of modules found in Annexes to the CE Directive

Module	*Title*	*Key points*
Module G	EC Declaration of Conformity	The manufacturer produces technical documentation that would allow assessment of the product to CE requirements. Involvement of the Notified Body is limited to recording the registration of the manufacturer and investigating any reported non-compliance.
Module H	Full Quality Assurance (ISO 9001:1994)	The manufacturer applies an approved QMS that addresses quality for the design, manufacture, and final inspection of a product. Involvement of the Notified Body includes approval/testing of the QMS and Products from the design stage right through to completion and post-production follow up.

Table 5.2 continued

5.10 Manufacturers

A manufacturer, in the meaning of New Approach, is the person who is responsible for designing and manufacturing a product with a view of placing it on the EU market on his own behalf.

The manufacturer may use finished products, ready-made parts or components, or he may subcontract his tasks. However, he must always retain the overall control and have the necessary competence to take the responsibility for the product.

The manufacturer has an obligation to ensure that a product intended to be placed on the EU market is designed and manufactured, and is conformity assessed to the Essential Requirements in accordance with the provisions of the applicable New Approach Directive(s).

5.11 Manufacturer's tasks under each module

Table 5.6 indicates the manufacturer's (and/or his representative's) tasks under each Module and its variant.

Module	*Title*	*Summary of procedure*
Module A	Internal Production Control	• Manufacturer writes product conformity details sufficient to allow the understanding/assessment of design, manufacture and performance of the product • Manufacturer maintains records for 10 years*
Module B	EC-Type Examination	• Manufacturer writes product conformity details sufficient to allow the understanding of design, manufacture and performance of the product • Notified Body assesses documentation • Notified Body issues EC-Type Examination certificate if product conforms • Manufacturer affixes CE Marking • Inform Notified Body of any significant changes to product • Manufacturer maintains records for 10 years*
Module D	Production Quality Assurance (IS0 9002:1994)	• Manufacturer affixes CE Marking • Manufacturer writes Declaration of Conformity • Manufacturer writes and applies QMS including a written undertaking to fulfil the obligations of the Quality Management System • Manufacturer lodges application for the exam of the QMS including a written undertaking to fulfil the obligations of the QMS • Manufacturer also commits to setting up and applying a systematic procedure for the review of experience gained in the post-production stage • Notified Body audits QMS and visits the manufacturer's premises and certifies whether the QMS conforms to CE requirements • Changes to the approved QMS must be re-certified by the Notified Body

Table 5.3 Summary of conformance procedures found in the Annexes to the CE Directive

Module	*Title*	*Summary of procedure*
		• Notified Body carries out periodic inspections (may be unannounced) • Manufacturer maintains records for 10 years*
Module E	Product Quality Assurance (ISO 9003:1994)	• Manufacturer affixes CE Marking • Manufacturer writes Declaration of Conformity • Manufacturer writes and applies QMS including a written undertaking to fulfil the obligations of the Quality Management System • Manufacturer lodges application for the exam of the QMS including a written undertaking to fulfil the obligations of the QMS • Manufacturer also commits to setting up and applying a systematic procedure for the review of experience gained in the post-production stage • Notified Body audits QMS and visits the manufacturer's premises and certifies whether the QMS conforms to CE requirements • Changes to the approved QMS must be re-certified by the Notified Body • Notified Body carries out periodic inspections (may be unannounced) • Manufacturer maintains records for 10 years*
Module F	Product Verification	• Manufacturer produces documentation supporting the design, manufacture and performance of a product • Manufacturer affixes CE Marking • Manufacturer writes Declaration of Conformity • Set up and apply a systematic procedure for the review of experience gained in the post-production stage • Notified Body carries out inspection and testing to verify that CE requirements are met either by batch sampling or checks on every product

Table 5.3 continued

Module	*Title*	*Summary of procedure*
		• Notified Body fixes ID number to each approved product and draws up written certificate of conformity • Manufacturer maintains records for 10 years*
Module G	EC Declaration of Conformity	• Manufacturer writes product conformity details sufficient to allow the understanding/assessment of design, manufacture and performance of the product • Set up and apply a systematic procedure for the review of experience gained in the post-production stage
Module H	Full Quality Assurance (ISO 9001:1994)	• Manufacturer writes Declaration of Conformity • Manufacturer affixes CE Marking • Manufacturer writes and applies QMS including a written undertaking to fulfil the obligations of the QMS • Manufacturer also commits to setting up and applying a systematic procedure for the review of experience gained in the post-production stage • Notified Body audits Quality Management System • Manufacturer applies for an examination of the design dossier of a given product • Notified Body issues EC-design exam certificate if product design conforms • Changes to approved designs must be re-certified by the Notified Body • Notified Body carries out periodic inspections (may be unannounced) • Manufacturer maintains records for 10 years*

* Depends on Directive

Table 5.3 continued

Title	*Module A*	*Module B*	*Module C*	*Module D*	*Module E*	*Module F*	*Module G*	*Module H*
CE Marking	X	X	X	X	X	X	X	X
EC Declaration of Conformity	X	X	X	X	X	X	X	X
EC Certificate of Conformity (e.g. via the 'TCF Route' for EMC Products)	X							
EC-Type Examination		X	X	X	X	X	X	X
Product Quality Assurance					X			
Production Quality Assurance				X				
Full Quality Assurance								X
Product Verification						X		
Technical File	X	X	X	X	X	X	X	X
Unit Verification							X	
Internal Production Control	X							
Conformity to Type			X					

Table 5.4 Conformance Requirements for each Module

Title	*73/23/ EEC*	*87/404/ EEC*	*88/378/ EEC*	*89/106/ EEC*	*89/336/ EEC*	*89/686/ EEC*	*90/384/ EEC*	*90/385/ EEC*	*90/396/ EEC*	*92/42/ EEC*	*93/15/ EEC*	*93/42/ EEC*	*94/9/ EC*	*94/25/ EC*	*95/16/ EC*	*COM(95) 130*	*96/57/ EC*	*97/23/ EC*	*98/37/ EC*	*99/5/ EC*
CE Marking	X	X	X	X	X	X	X	X	X	X	X	X	X	X	X	X	X	X	X	X
EC Declaration of Conformity	X	X	X	X	X	X	X	X	X	X	X	X	X	X	X	X	X	X	X	X
EC Certificate of Conformity – Module C		X																		
EC-Type Examination – Module B		X			X	X	X	X	X	X	X	X	X	X	X			X	X	
Product Quality Assurance – Module E (ISO 9003)						X				X	X	X	X		X			X	X	
Production Quality Assurance – Module D (ISO 9002)						X				X	X	X	X	X	X			X		
Full Quality Assurance – Module H (ISO 9001)									X			X			X			X	X	X
Product Verification – Module F							X	X	X		X	X	X					X		
Technical File	X	X	X	X	X	X	X	X	X	X	X	X	X	X	X	X	X	X	X	X
Unit verification – Module G							X		X		X		X					X		
Internal Production Control – Module AI	X										X		X	X	X		X	X	X	X
Conformity to Type – Module C											X		X	X				X	X	

Table 5.5 Conformity Procedures

Table 5.6 Manufacturer's tasks under each module

Module	*Manufacturer*	*Manufacturer or his authorised representative*
A	Establishes technical documentation as regards the design and manufacture of the product Takes all measures necessary to ensure that the manufacturing process assures compliance of the products with the technical documentation and with the applicable requirements (i.e. operates a quality system)	Ensures and declares that the products concerned satisfy the Essential Requirements Affixes the CE Marking to each product Draws up a Declaration of Conformity Keeps a copy of the Declaration of Conformity and the technical documentation at the disposal of the surveillance authorities
Aa1	In addition to the responsibilities as in module A: • carries out one or more tests for each product manufactured	In addition to the responsibilities as in module A: • applies for supervision of testing • affixes the Notified Body's identification number to follow the CE marking, if the Notified Body intervened during the production stage
Aa2	As in module A	In addition to the responsibilities as in module A: • applies for product checks • affixes the Notified Body's identification number to follow the CE marking
B	Establishes technical documentation as regards the design of the product	Applies for the EC-type examination Places at the disposal of the Notified Body one (or more) specimen(s), which is (are) representative of the production envisaged

Module	*Manufacturer*	*Manufacturer or his authorised representative*
B continued		Informs the Notified Body of all modifications to the approved product Keeps the technical documentation, including a copy of the EC-type examination certificate, at the disposal of the surveillance authorities
C	Establishes technical documentation as regards the manufacture of the product Takes all measures necessary to ensure that the manufacturing process assures compliance of the products with the type as described in the EC-type examination certificate and with the applicable requirements (i.e. operates a quality system)	Ensures and declares that the products concerned are in conformity with the EC-type examination certificate and satisfy the applicable requirements Affixes the CE Marking to each product Draws up a Declaration of Conformity Keeps relevant technical information and a copy of the Declaration of Conformity at the disposal of the surveillance authorities
Cbis1	As in module Aa1, in addition to the responsibilities as in module C	As in module Aa1, in addition to the responsibilities as in module C
Cbis2	As in module Aa2, in addition to the responsibilities as in module C	As in module Aa2, in addition to the responsibilities as in module C
D	Operates an approved quality system for production, final product inspection and testing, which includes the drawing up of technical documentation (i.e. relevant information for the product category envisaged, documentation concerning the quality system and its updating,	Applies for the assessment of the quality system for the products concerned Ensures and declares that the products concerned are in accordance with the EC-type examination certificate and satisfy the applicable requirements

D continued	technical documentation of the approved type, a copy of the EC-type examination certificate, and the decisions and reports from the Notified Body) Undertakes to fulfil the obligations arising out of the approved quality system and upholds it so that it remains adequate and efficient Supports the action carried out by the Notified Body for surveillance purposes	Affixes the CE Marking to each product Affixes the Notified Body's identification number to follow the CE marking Draws up a Declaration of Conformity Informs the Notified Body of any intended updating of the quality system Keeps a copy of the Declaration of Conformity at the disposal of the surveillance authorities Ensures that he is able to supply the relevant technical documentation to the surveillance authority
Dbis	Establishes technical documentation as regards the design of the product Operates an approved quality system for production, final product inspection and testing, which includes the drawing up of technical documentation (i.e. relevant information for the product category envisaged, documentation concerning the quality system and its updating, and the decisions and reports from the Notified Body) Undertakes to fulfil the obligations arising out of the approved quality system and upholds it so that it remains adequate and efficient Supports the action carried out by the Notified Body for surveillance purposes	Applies for the assessment of the quality system for the products concerned Ensures and declares that the products concerned satisfy the requirements Affixes the CE Marking to each product Affixes the Notified Body's identification number to follow the CE Marking Draws up a Declaration of Conformity Informs the Notified Body of any intended updating of the quality system Keeps a copy of the Declaration of Conformity at the disposal of the surveillance authorities Ensures that he is able to supply the relevant technical documentation to the surveillance authority

Module	*Manufacturer*	*Manufacturer or his authorised representative*
E	As in module D, but operates an approved quality system for final product inspection and testing	As in module D
Ebis	As in module Dbis, but operates an approved quality system for final product inspection and testing	As in module Dbis
F	Establishes technical documentation as regards the manufacture of the product Takes all measures necessary to ensure that the manufacturing process assures conformity of the products with the type as described in the EC-type examination certificate and with the applicable requirements (i.e. operates a quality system) Where the statistical verification is used: presents his products in the form of homogeneous lots and takes all measures necessary in order that the manufacturing process ensures the homogeneity of each lot produced	Applies for certification of conformity Checks and attests that the products are in conformity with the type as described in the EC-type examination certificate and satisfy the applicable requirements Affixes the CE Marking to each product Affixes the Notified Body's identification number to follow the CE Marking Draws up a Declaration of Conformity Keeps a copy of the Declaration of Conformity and the technical documentation at the disposal of the surveillance authorities Ensures that he is able to supply the notified body's certificates of conformity to the surveillance authority on request
Fbis	Establishes technical documentation as regards the design and manufacture of the product	Applies for certification of conformity Checks and attests that the products satisfy the applicable requirements

Fbis continued	Takes all measures necessary to ensure that the manufacturing process assures conformity of the products with the applicable requirements (i.e. operates a quality system) Where the statistical verification is used: presents his products in the form of homogeneous lots and takes all measures necessary in order that the manufacturing process assures the homogeneity of each lot produced	Affixes the CE Marking to each product Affixes the Notified Body's identification number to follow the CE Marking Draws up a Declaration of Conformity Ensures that he is able to supply the Notified Body's certificates of conformity on request Keeps a copy of the Declaration of Conformity and the technical documentation at the disposal of the surveillance authorities
G	Establishes technical documentation as regards the design and manufacture of the product	Applies for certification of conformity Ensures and declares that the product concerned conforms to the applicable requirements Affixes the CE Marking to each product Affixes the Notified Body's identification number to follow the CE Marking Draws up a Declaration of Conformity Keeps a copy of the Declaration of Conformity and the technical documentation at the disposal of the surveillance authorities

Module	*Manufacturer*	*Manufacturer or his authorised representative*
H	Operates an approved quality system for design, manufacture, final product inspection and testing, which includes the drawing up of technical documentation (i.e. relevant information for the design, the product category envisaged, documentation concerning the quality system and its updating, and the decisions and reports from a Notified Body) Undertakes to fulfil the obligations arising out of the approved quality system and upholds it so that it remains adequate and efficient Supports the action carried out by the Notified Body for surveillance purposes	As in module Dbis
Hbis	As in module H	In addition to responsibilities as in module Dbis: Applies for examination of the design Informs the Notified Body of any modification to the approved design

Table 5.6 Manufacturer's tasks under each module

5.12 Basic requirements of manufacturers of industrial products

The following list is a summary of the seven basic requirements that manufacturers of industrial products are expected to comply with:

1 **Safety**. An industrial product must be designed and manufactured so that it will not compromise the safety or well being of the users (i.e. those people entrusted with fitting, maintaining and using the industrial product). Should there be any risk involved with using the product, then this risk must be weighed against the benefits to the user. The product should **only** be used if the risk is acceptable.
2 **Principles of Design**. The design and construction of an industrial product must conform to internationally agreed safety principles and the accepted state of the art.
3 **Risk**. When designing and manufacturing an industrial product the following principles must be applied:
   - risk shall be eliminated or reduced as far as possible;
   - adequate measures to protect against any risks that cannot be eliminated shall be taken;
   - the user shall be informed of any shortcomings in the protection measures adopted.

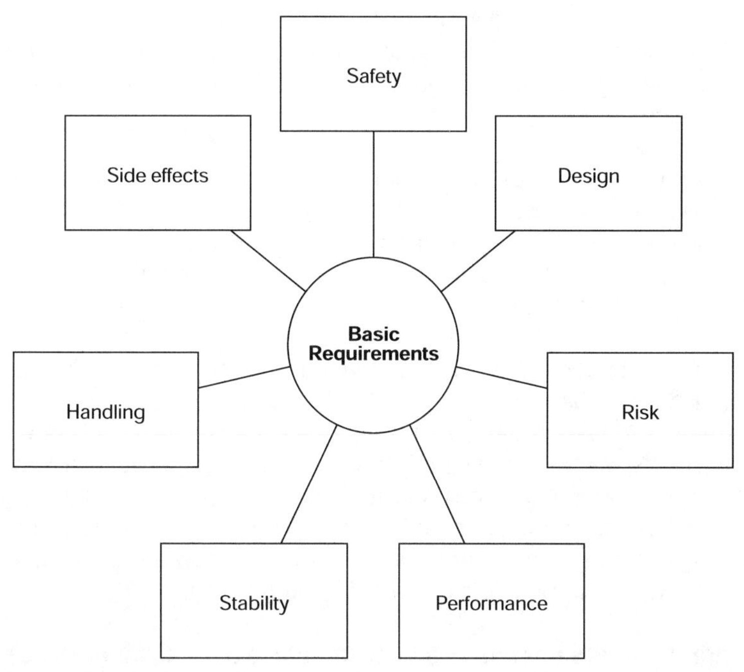

Figure 5.3 Basic requirements of manufacturers of industrial products

4 **Performance**. The industrial product must perform to the manufacturer's requirements (i.e. the standards set by the manufacturer) and be fit for purpose. In other words it should be capable of surviving the type of environment for which it was designed.
5 **Stability**. The characteristics of the industrial product must not alter during its expected lifetime.
6 **Handling**. The industrial product must be designed, manufactured and packaged so that it can withstand the effects of storage, transportation and use.
7 **Side effects**. Any undesirable side effect must constitute an acceptable risk when weighed against the intended performance of the industrial product.

Depending upon the type of product the above requirements are supplemented and enlarged upon in the various Directives. However, these seven basic principles are true for all industrial products.

One other major requirement that is worthy of note at this time is the need for a manufacturer to **prove** that they are complying with these requirements. This may be achieved by compiling and working to an auditable Quality Management System (QMS). Further details of what is required and how to complete the documentation can be found in two of my other books, *ISO 9000 for Smaller Companies* or *MDD Compliance using Quality Management Techniques*, both of which are available from stingray@herne.demon.co.uk.

The Commission's *Guide to the Implementation of Community Harmonisation Directives Based on the New Approach and Global Approach* states:

> Manufacturers are responsible for ensuring that the products they place on the market meet all the relevant regulations. Where these regulations do not require mandatory certification, manufacturers often seek voluntary certification to assure themselves that their products do meet the requirements set by law.

From a manufacturer's point of view, in addition to the standards, a manufacturer must also check the Essential Requirements of the Directives and take into account the generally acknowledged state of the art. Products must be designed and manufactured in such a way that, when used under the conditions and intended purposes, they will not compromise the health and safety of persons or the environment. In selecting the most appropriate design solutions, the manufacturer shall apply the following principles in order of preference:

1 **Inherently safe design**. Eliminate or reduce risks as far as possible by construction.

2 **Guarding and safety components**. When an inherently safe design is not possible, use proper protective measures such as guarding, safety components, alarms, in relation to risks that cannot be eliminated.
3 **Warnings and instructions**. Inform users of residual risks due to shortcomings of design. This method is to be employed only as a last resort and when (1) or (2) is not possible.

When choosing standards, a manufacturer has to consider the following points:

- **Type of product**: the scope of the standard needs to be checked to determine if it refers to the product.
- **Environment**: the standard has to deal with the conditions of usage.
- **User**: the skill and protection of the operator and service person has to be considered.
- **Other considerations**: other standards may be applicable.

5.13 Registration of manufacturers

Any manufacturer who places products on the market under his own name must be registered with the Competent Authority.

All the procedures detailed in Table 5.6 have one thing in common, that being the need to have some form of auditable documentation to prove that a manufacturer has adequate control over quality. In some cases the Directives call for full documentation on management structure, design, manufacturing and final inspection of a product; in other cases they simply ask for technical details sufficient to allow the understanding and assessment of product conformity.

In most cases the Directives state that the documentation is to be presented in the form of a **Quality Management System**. In the other Directives no such statement is made, but in reality the requirements of these Directives should also be met by setting up and implementing Quality Management Systems.

5.14 EC Verification

EC Verification is the procedure whereby the manufacturer, or his authorised representative, ensures and declares that the industrial products that he is producing conform to the type described in the EC-Type Examination certificate and meet the applicable CE requirements.

Whilst Modules A, B, C, F and G do not themselves ask for a QMS, however, the format of the conformity procedure is such that its documentary requirements are best presented in the form of a Quality Management System.

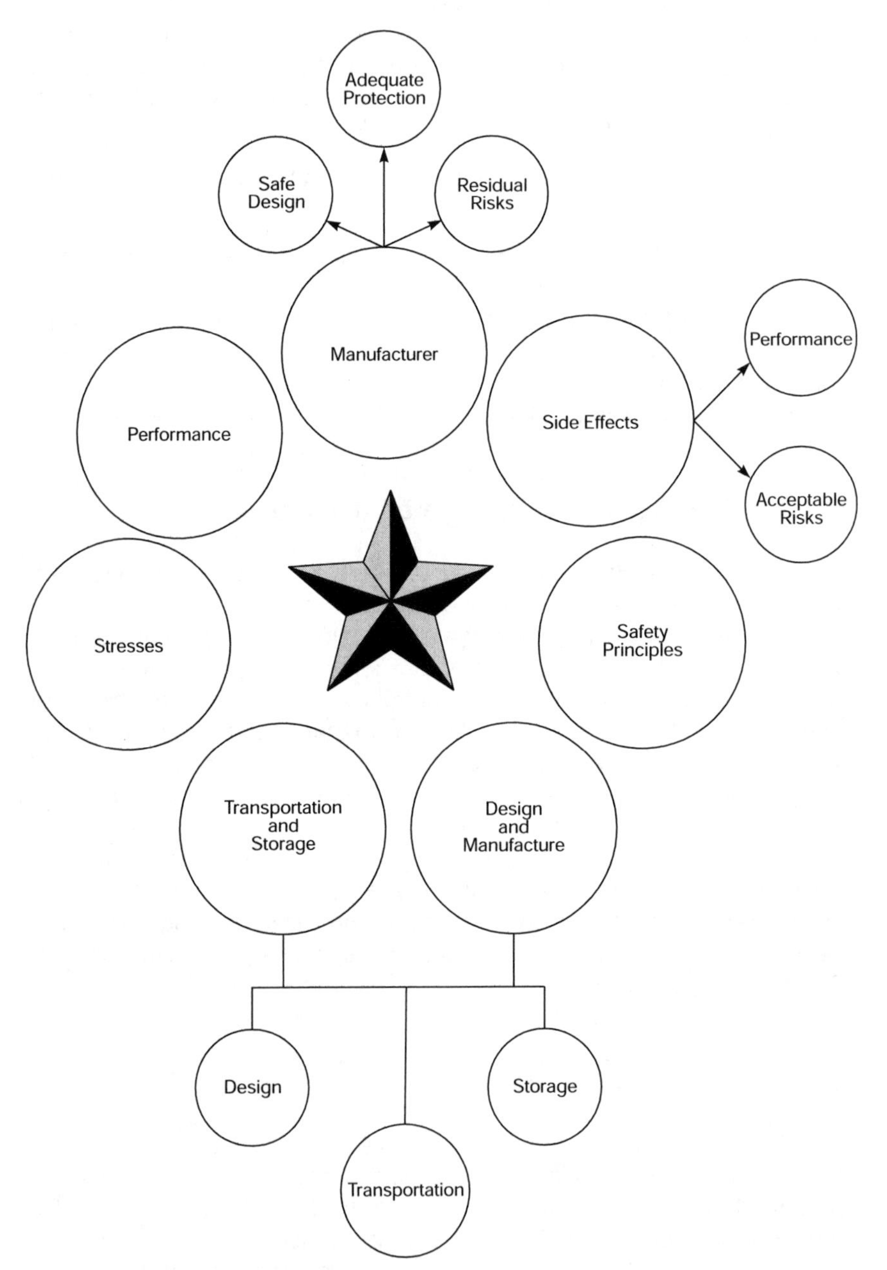

Figure 5.4 Essential (manufacturing) requirements

The following are considered as essential requirements for manufactured goods to achieve compliance with the appropriate Directives and CE Marking.

1 Industrial products must be designed and manufactured so that when they are used under normal conditions (i.e. the purpose for which they were originally intended), they will not compromise the safety or health of users.
2 The design and construction of an industrial product must conform to safety principles and the accepted state of the art.
3 The manufacturer shall:
   - eliminate and/or reduce risks as far as possible (i.e. provide an inherently safe design and construction);
   - take adequate protection measures (including alarms if necessary) for any risks that cannot be completely eliminated;
   - inform users of any residual risks due to any shortcomings of the protection measures that they have adopted.
4 Industrial products shall achieve the performances intended by the manufacturer.
5 The characteristics and performance of industrial products shall not be affected (in any way) so as to compromise the safety of the users during the lifetime of the product – when that product is subjected to the stresses found during normal use.
6 Industrial products shall be designed, manufactured and packed so that they can be transported and stored without affecting the characteristics and performance of the industrial product.
7 Any side effect caused by an industrial product shall not constitute an unacceptable risk to the (original) performances intended.

5.15 Design and Construction Requirements

5.15.1 Chemical, physical and biological properties

Industrial products shall be designed and manufactured in accordance with the Essential Requirements stated in the relevant Directive(s). The product's chemical, physical and biological properties shall be considered with particular attention being paid to:

- choice of materials used (e.g. toxicity and flammability etc.);
- compatibility between the materials used and biological tissues, cells and body fluids;
- the risk posed by contaminants and residues to persons (and patients) involved in the transportation storage and use of the industrial product;
- being able to use an industrial product safely, no matter what materials, substances and gases that it may come into contact with;

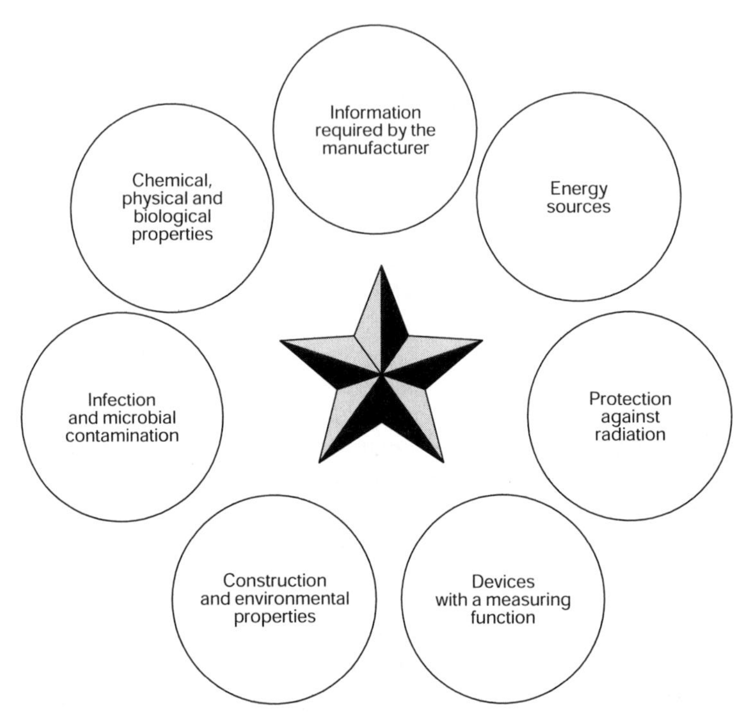

Figure 5.5 Design and construction requirements

- the compatibility of an industrial product (which is intended to administer medicinal products) with the provisions and restrictions of those medicinal products;
- the verification of the usefulness of any substance that is an integral part of an industrial product. In particular:
  - any substances leaking from the product shall be minimised;
  - the risks associated with the possible unintentional ingress of substances into the product shall be reduced as much as possible.

5.15.2 Construction and environmental properties

If the industrial product is going to be used in conjunction with another product or equipment, the whole combination (including any connection system) must be safe and the industrial product must not affect the specified performances of itself or any other associated product. Any restrictions regarding use etc. shall be indicated on the label or in the instruction manual.

Products shall be designed and manufactured in such a way as to remove or minimise (as far as possible):

- the risk of injury (caused by their physical, dimensional and/or ergonomic features);
- risks connected with reasonably foreseeable environmental conditions, such as magnetic fields, external electrical influences, electrostatic discharge, pressure, temperature or variations in pressure and acceleration;
- the risks of mutual interference with other products;
- risks arising where maintenance or calibration are not possible (as with implants) from ageing of materials;
- loss of accuracy of any measuring or control mechanism through the ageing of materials;
- the risks of fire or explosion during normal use and in single fault condition.

5.15.3 Products with a measuring function

The manufacturing process for products with a measuring function shall be designed in such a way as to:

- take into account the intended purpose of the industrial product;
- provide sufficient accuracy and stability;
- ensure that the measurement, monitoring and display scale has been designed in line with current and applicable ergonomic principles;
- conform to the legal requirements and provisions of Council Directive 80/181/EEC as amended by 89/617/EEC.

Note: In all circumstances the manufacturer must indicate the limits of accuracy.

5.15.4 Protection against radiation

Whilst not actually restricting the use of that product, all industrial products shall be designed and manufactured so that the exposure of patients, users and other persons to radiation is reduced as far as possible.

Note: This is particularly relevant in therapeutic and diagnostic applications.

5.15.4.1 Intended radiation

If industrial products are designed to emit hazardous levels of radiation, it must be possible for the user to control the emissions.

Where products are intended to emit potentially hazardous, visible and/ or invisible radiation, they must be fitted, where practicable, with visual displays and/or audible warnings of such emissions.

5.15.4.2 Unintended radiation

Exposure of users to the emission of unintended, stray or scattered radiation shall be reduced as far as possible.

5.15.4.3 Instructions

Operating instructions for products emitting radiation shall provide detailed information regarding the nature of the emitted radiation, the means of protecting the patient, the methods for avoiding misuse and eliminating inherent installation risks.

5.15.4.4 Ionising radiation

Products intended to emit ionising radiation shall ensure that the quantity, geometry and quality of the radiation being emitted can be varied and controlled.

Products emitting ionising radiation that are intended for diagnostic radiology shall minimise the amount of radiation that the patient and user are exposed to.

Products emitting ionising radiation that are intended for therapeutic radiology shall ensure reliable monitoring and control of the delivered dose, the beam type and energy.

5.15.5 Requirements for industrial products connected to or equipped with an energy source

Industrial products which are designed to be connected to an electrical source shall be designed so that:

- products incorporating electronic programmable systems ensure the continued reliability and performance of these systems according to the intended use;
- products where the safety of the patients depends on an internal power supply are equipped with a means of determining the state of the power supply;
- products where the safety of the patients depends on an external power supply must include an alarm system to indicate a power failure;
- products intended to monitor one or more of the patient's clinical parameters must be equipped with an appropriate alarm system that is capable of alerting the user of any situation which could lead to the possible death or severe deterioration of the patient's parameters;
- the risks of creating electromagnetic fields, which could impair the operation of other products or equipment, are minimised.

5.15.5.1 Protection against electrical risks

Products must be manufactured in such a way as to avoid, as far as possible, the risk of accidental electric shocks during normal use and in single fault condition.

5.15.5.2 Protection against mechanical and thermal risks

The manufacture and use of an industrial product shall ensure that:

- the patient and/or user are protected against mechanical risks connected with, for example, resistance, stability and moving parts;
- the level of risk arising from vibration generated by the products is reduced as low as possible;
- there is a minimum possible level of risk arising from emitted noise to the patient and the user;
- any terminals and connectors to the electricity, gas, hydraulic or pneumatic energy supplies (which the user has to handle) have been designed and constructed in such a way as to minimise all possible risks;
- all accessible parts of the product (excluding the parts or areas intended to supply heat or reach given temperatures) and their surroundings do not attain potentially dangerous temperatures under normal use.

5.15.6 Manufacturer's instructions

Where appropriate, instructions for use must contain the following:

- The manufacturer's label indicating:
  - name (or trade name) and address of the manufacturer;
  - details to identify the product and (where appropriate) the contents of the packaging;
  - where appropriate, the serial number of the product;
  - any special storage and/or handling conditions;
  - any special operating instructions;
  - any warnings and/or precautions to take;
  - year of manufacture (**Note**: This indication may be included in the batch or serial number).
- the required performance and details of any undesirable side-effects;
- if the product is to be installed or connected to another industrial product/equipment, sufficient details in order to obtain a safe combination;
- all the information required to verify that the product is properly installed and can operate correctly and safely, plus full details (e.g. nature and frequency) of any maintenance and calibration restrictions.

5.15.7 Manufacturer's cleaning and sterilisation instructions

Where products are supplied with the intention that they must be sterilised before use, the instructions for cleaning and sterilisation shall include:

- details of any further treatment or handling required before the product can be used (e.g. sterilisation);
- in the case of products emitting radiation for medical purposes, details of the nature, type, intensity and distribution of this radiation.

5.15.8 Authorised representatives

The manufacturer may appoint any natural or legal person to act on his behalf as an authorised representative. These people are not defined in the New Approach Directives, with the exemption of the Directive on in-vitro diagnostic medical products. For the purposes of New Approach Directives the authorised representative must be established inside the EU.

The authorised representative is explicitly designated by the manufacturer, and he may be addressed by the authorities of the Member States instead of the manufacturer with regard to the latter's obligations under the New Approach Directive in question.

The manufacturer remains generally responsible for actions carried out by an authorised representative on his behalf.

5.16 Importer/person responsible for placing the product on the market

An importer (i.e. the person responsible for placing on the market) – in the meaning of New Approach Directives – is any natural or legal person established in the EU who places a product from a third country on the EU market.

The importer must ensure that he is able to provide the market surveillance authority with the necessary information regarding the product, where the manufacturer is not established in the EU, and he has no authorised representative in the EU. The natural or legal person who imports a product into the EU may, in some situations, be considered as the person who must assume the responsibilities placed on the manufacturer according to the applicable New Approach Directives.

5.17 Distributor

Provisions regarding distribution are in general not included in New Approach Directives.

A distributor is to be considered as any natural or legal person in the supply chain who takes subsequent commercial actions after the product has been placed on the EU market. The distributor must act with due care in order not to place clearly non-compliant products on the EU market. He must also be capable of demonstrating this to the National Surveillance Authority.

5.18 Assembler and Installer

The installer and assembler of a product, which is already placed on the market, should take necessary measures to ensure that it still complies with the essential requirements at the moment of first use within the EU. This applies to products where the Directive in question covers putting into service, and where such manipulations may have an impact on the compliance of the product.

5.19 User

New Approach Directives do not lay down obligations for users with the exception that according to the Directive on pressure equipment Member States may, under certain conditions, authorise in their territory the placing on the market, and the putting into service by users.

EU legislation concerning the health and safety of the workplace has an impact on the maintenance and use of products covered by New Approach Directives that are used in the workplace.

Annex A: Glossary

Acceptable quality level: A measure of the number of failures that a production process is allowed. Usually expressed as a percentage.

Acceptance: Agreement to take a product or service as offered.

Accessory: An article which whilst not being a device is intended specifically by its manufacturer to be used together with a device to enable it to be used in accordance with the use of the device intended by the manufacturer of the device.

Active implantable medical device: Any active medical device which is intended to be totally or partially introduced, surgically or medically, into the human body or by medical intervention into a natural orifice, and which is intended to remain after the procedure.

Active medical device: Any medical device relying for its functioning on a source of electrical energy or any source of power other than that directly generated by the human body or gravity.

Apparatus: All electrical and electronic appliances together with equipment and installations containing electrical and/or electronic components.

Approval: The decision taken to allow envisaged transfers of explosives within the EU.

Assemblies: Several pieces of pressure equipment assembled by a manufacturer to constitute an integrated and functional whole.

Basic specification: A specification that is applicable to all components or to a large group of components.

Batch: A quantity of some commodity that has been manufactured or produced under uniform conditions.

Boiler: The combined boiler body-burner unit, designed to transmit to water the heat released from burning.

Boiler, appliance:

- the boiler body designed to have a burner fitted;
- the burner designed to be fitted to a boiler body.

Boiler, back: A boiler designed to supply a central-heating system and to be installed in a fireplace recess as part of a back boiler/gas fire combination.

Boiler, gas condensing: A boiler designed to condense permanently a large part of the water vapour contained in the combustion gases.

Boiler, low temperature: A boiler which can work continuously with a water supply temperature of 35 to 40°C, possibly producing condensation in certain circumstances, including condensing boilers using liquid fuel.

Boiler, standard: A boiler for which the average water temperature can be restricted by design.

Boiler, water average temperature: The average of the water temperatures at the entry and exit of the boiler.

Boilers, effective rated output: The maximum calorific output laid down and guaranteed by the manufacturer as being deliverable during continuous operation while complying with the useful efficiency indicated by the manufacturer.

Boilers, part load: The ratio between the effective output of a boiler operating intermittently or at an output lower than the effective rated output and the same effective rated output.

Boilers, useful efficiency: The ratio between the heat output transmitted to the boiler water and the product of the net calorific value at constant fuel pressure and the consumption expressed as a quantity of fuel per unit time.

Bonded store: A secure place in which only supplies that have been accepted as satisfactory by the inspection staff are held.

Calibration: The operation that is required to determine the accuracy of measuring and test equipment.

Capability approval: Approval that is granted to a manufacturer when they have demonstrated that their declared design capability, manufacturing processes and quality control meets the requirements of the relevant generic specification.

Capability qualifying components: A group of components and/or test pieces which are collectively used to demonstrate that the declared capability meets the requirements that have been specified in the generic specification.

CE Marking: CE Marking is the manufacturer's or supplier's self-declaration symbol to indicate that the product has undergone all the necessary evaluation procedures and is in conformity with the minimum requirements of the relevant European directives.

CEN (European Committee for Standardisation): European equivalent of ISO.

CENELEC (European Committee for Electrotechnical Standardisation) Certification Body: An impartial body who have the necessary competence and reliability to operate a certification scheme.

Certificate of Conformity: A document stating that, at the time of the assessment, the product or service met the stated requirements.

Certification Body: An impartial body who have the necessary competence and reliability to operate a certification scheme.

Certification system: A system for carrying out conformity certification.

Certified test record: A summary of the results of specified tests that have been carried out on components that were released over the last six month production period.

Characteristic: A property that helps to distinguish between items of a given population.

Company: Term used primarily to refer to a business first party, the purpose of which is to supply a product or service.

Competent body: Any body which meets the criteria listed in Annex II and is recognised as such.

Compliance: The fulfilment of a Quality Management System or quality procedure of specified requirements.

Components: Any item essential to the safe functioning of equipment and protective systems but with no autonomous function.

Concession: Written authorisation to use or release a quantity of material, components or stores already produced but which do not conform to the specified requirements.

Configuration: The complete technical description that is required to make, test, equip, install, operate, maintain and logistically support a product.

Conformance: The fulfilment of a product or service of specified requirements.

Consignment: Products (or goods) that are issued or received as one delivery and covered by one set of documents.

Construction product: Any product which is produced for incorporation in a permanent manner in construction works, including both buildings and civil engineering works.

Contract: Agreed requirements between a supplier and customer transmitted by any means.

Contractor assessment: The formal examination by a National Quality Assurance Authority to determine the ability of a contractor or potential contractor to meet requirements.

Corrective maintenance: The maintenance that is carried out after a failure has occurred and which is intended to restore the item to a state where it can perform its original function.

Custom-made device: Any active implantable medical device specifically made in accordance with a medical specialist's written prescription which gives, under his responsibility, specific design characteristics and is intended to be used only for an individual named patient.

Customer: Ultimate consumer, user, client, beneficiary or second party.

Customer complaint: Any written, electronic, or oral communication that alleges deficiencies related to the identity, quality, durability, reliability, safety or performance of a device that has been placed on the market.

Dealer: Any natural or legal person whose occupation consists wholly or partly in the manufacture, trade, exchange, hiring out, repair or conversion of fire arms and ammunition.

Delivery lot: A quantity of components that are delivered at the same time.

Design authority: The approved firm, establishment or branch representative responsible for the detailed design of material to approved specifications and authorised to sign a certificate of design, or to certify sealed drawings.

Design capability: The ability of a manufacturer to translate a customer requirement into a component that can be manufactured by their particular technology.

Design review: A formal documented, comprehensive and systematic examination of a design to evaluate the design requirements and the capability of the design to meet these requirements and to identify problems and propose solutions.

Detail specification: A specification which is derived from a generic or sectional specification, which covers and describes a particular component or a recognised range of components.

Development: Life cycle process that comprises the activities or requirements analysis, design, coding, integration, testing, installation and support for acceptance of software products.

Deviation permit: Written authorisation, prior to production or before provision of a service, to depart from specified requirements for a specified quantity or for a specified time.

Device intended for clinical investigation: Any active implantable medical device intended for use by a specialist doctor when conducting investigations in an adequate human clinical environment.

Device master record: A compilation of records containing the design, specification, complete manufacturing procedures, quality assurance requirements and labels and labelling of a finished device.

Device used for in-vitro diagnosis: Device which is a reagent, reagent product, kit, instrument, equipment or system, whether used alone or in combination, intended by the manufacturer to be used in vitro for the examination of samples derived from the human body with a view to providing information on the physiological state of health or disease, or congenital abnormality thereof.

Direct surveillance: Surveillance carried out on premises that come under the direct control of the chief inspector by reason of his appointment.

Distributor: An organisation that is contractually authorised by one or more manufacturers to store, repack and sell completely finished components from these manufacturers.

EC-Type Examination Certificate: A document in which a notified body referred to in Article 10 (6) certifies that the type of equipment examined complies with the provisions of this Directive which concern it.

Economic quality: The economic level of quality at which the cost of obtaining higher quality would exceed the benefits of the improved quality.

Effect: The non-fulfilment of intended usage requirements.

Electromagnetic compatibility: The ability of a device, unit of equipment or system to function satisfactorily in its electromagnetic environment without introducing intolerable electromagnetic disturbances to anything in that environment.

Electromagnetic disturbance: Any electromagnetic phenomenon which may degrade the performance of a device, unit of equipment or system.

An electromagnetic disturbance may be electromagnetic noise, an unwanted signal or a change in the propagation medium itself.

Electronic component: A device that is part of an electronic circuit and that has a distinctive function in that electronic circuit.

Environment: All of the external physical conditions that may influence the performance of a product or service.

Environmental condition: The characteristics (such as humidity, pressure, vibration etc.) of the environment in which the product is operating.

Environmental stress: The stress to which a product is exposed that is solely due to its presence in an environment.

Equipment: Machines, apparatus, fixed or mobile devices, control components and instrumentation thereof and detection or prevention systems which, separately or jointly, are intended for the generation, transfer, storage, measurement, control and conversion of energy for the processing of material and which are capable of causing an explosion through their own potential sources of ignition.

Equipment class: A class identifying particular types of apparatus which under this Directive are considered similar and those interfaces for which the apparatus is designed. Apparatus may belong to more than one equipment class.

European approval for materials: A technical document defining the characteristics of materials intended for repeated use in the manufacture of pressure equipment which are not covered by any harmonised standard.

Evaluation: The systematic evaluation of the effectiveness of a contractor's Quality Management System.

Explosive atmospheres: Mixture with air, under atmospheric conditions, of flammable substances in the form of gases, vapours, mists or dusts in which, after ignition has occurred, combustion spreads to the entire unburned mixture.

Explosives: The materials and articles considered to be such in the United Nations recommendations on the transport of dangerous goods and falling within Class 1 of those recommendations.

Facilities: The tools, materials, supplies, instruments, equipment and other resources that are available to manufacture a product or perform a service.

Fluids: Gases, liquids and vapours in pure phase as well as mixtures thereof. A fluid may contain a suspension of solids.

Full assessment: A degree of quality assessment that is higher than basic assessment level. Obtained by tighter inspection levels, tighter acceptable quality levels and more stringent tests.

Functional specification: A document that describes, in detail, the characteristics of the product with regard to its intended capability.

Functional stress: The stress to which a product is exposed that is solely due to its intended function.

Gaseous fuel: Any fuel which is in a gaseous state at a temperature of 15°C under a pressure of 1 bar.

Generic specification: A specification that is applicable to a family or sub-family of electronic components.

Grade: An indicator of category or rank related to features or characteristics that cover different sets of needs for products or services intended for the same functional use.

Graded standard: Defines the particular grade of an item of material or product for a particular application.

Harmful interference: Interference which endangers the functioning of a radionavigation service or of other safety services or which otherwise seriously degrades, obstructs or repeatedly interrupts a radio communications service operating in accordance with the applicable EU or national regulations.

Harmonised standard: A technical specification adopted by a recognised standards body under a mandate from the Commission in conformity with the procedures laid down in Directive 98/34/EC for the purpose of establishing a European requirement, compliance with which is not compulsory.

Immunity: The ability of a device, unit of equipment or system to perform without degradation of quality in the presence of an electromagnetic disturbance.

Independent test laboratory: An organisation that has the facilities and capability to carry out tests and measurements on electronic components in accordance with the relevant specification, and which does not form part of the manufacturing organisation producing these components.

Inshore craft: Craft designed for coastal waters, large bays, estuaries, lakes and rivers where conditions up to and including wind force 6 and significant wave heights up to and including 2 m.

Inspection: Activities such as measuring, examining, testing, gauging one or more characteristics of a product or service and comparing these with specified requirements to determine conformity.

Inspection by attributes: Inspection whereby certain characteristics of an item are assessed, without measurement, as either conforming or not conforming to the requirements of the product or service.

Inspection by variables: Inspection whereby certain characteristics of an item are evaluated against a numerical scale and are expressed as points along that scale.

Inspection lot: A collection of components or 'units' from which a sample is taken and inspected to determine conformance with the acceptability criteria.

Inspection system: The established management structure, responsibilities, methods, resources that together provide inspection.

Installer of a lift: The natural or legal person who takes responsibility for the design, manufacture, installation and placing on the market of the lift and who affixes the CE marking and draws up the EC declaration of conformity.

Intended purpose (as specified in Harmonised Directives): The use for which the device is intended according to the data supplied by the manufacturer on the labelling, in the instructions and/or in promotional materials.

Interchangeability: Versions of the same component type covered by a detail specification.

Interface:

- a network termination point, which is a physical connection point at which a user is provided with access to public telecommunications network, and/or
- an air interface specifying the radio path between radio equipment and their technical specifications.

International Organisation for Standardisation (ISO): Comprises the national standards bodies of more than 50 countries whose aim is to co-ordinate the international harmonisation of national standards.

Labelling: Written, printed or graphic matter:

- affixed to a device or any of its containers or wrappers, or
- accompanying a device,

related to identification, technical description and use of the device, but excluding shipping documents.

Lift: An appliance serving specific levels, having a car moving along guides which are rigid and inclined at an angle of more than 15° to the horizontal and intended for the transport of:

- persons;
- persons and goods;
- goods alone (if the car is accessible, that is to say, a person may enter it without difficulty, and fitted with controls situated inside the car or within reach of a person inside).

Long-term: A device normally intended for continuous use for more than 30 days.

Lot: A quantity of some commodity that has been manufactured or produced under uniform conditions.

Machinery:

- an assembly of linked parts or components, at least one of which moves, with the appropriate actuators, control and power circuits, etc., joined together for a specific application, in particular for the processing, treatment, moving or packaging of a material;
- an assembly of machines which, in order to achieve the same end, are arranged and controlled so that they function as an integral whole;
- interchangeable equipment modifying the function of a machine, which is placed on the market for the purpose of being assembled with a machine or a series of different machines or with a tractor by the operator himself in so far as this equipment is not a spare part or a tool.

Maintenance: The combination of technical and administrative actions that are taken to retain or restore an item to a state in which it can perform its stated function.

Manufacturer (as specified in the Medical Devices Directive): The natural or legal person with responsibility for the design, manufacture, packaging and labelling of a device before it is placed on the market under his own name, regardless of whether these operations are carried out by that person himself or on his behalf by a third party.

Manufacturer: An organisation which carries out or controls such stages in the manufacture of electronic components that enable it to accept responsibility for capability approval or qualification approval, inspection and release of electronic components.

Manufacturer of the safety components: The natural or legal person who takes responsibility for the design and manufacture of the safety components and who affixes the CE marking and draws up the EC declaration of conformity.

Manufacturer's specification: The specification that a manufacturer has agreed to meet at all costs and that has been accepted by the Design Authority as being sufficient to meet the User Requirement.

Material: A generic term covering equipment, stores, supplies and spares which form the subject of a contract.

Maximum allowable pressure (PS): The maximum pressure for which the equipment is designed, as specified by the manufacturer. It is defined at a location specified by the manufacturer. This must be the location of connection of protective and/or limiting devices or the top of equipment or if not appropriate any point specified.

Maximum/minimum allowable temperature (TS): The maximum/minimum temperatures for which the equipment is designed, as specified by the manufacturer.

May: This auxiliary verb indicates a course of action often followed by manufacturers and suppliers.

Medical device: Any instrument, apparatus, appliance, material or other article, whether used alone or in combination, including the software necessary for its proper application intended by the manufacturer to be used for human beings for the purpose of:

- diagnosis, prevention, monitoring, treatment or alleviation of disease;
- diagnosis, monitoring, treatment, alleviation of or compensation for an injury or handicap;
- investigation, replacement or modification of the anatomy or of a physiological process;
- control of conception;

and which does not achieve its principal intended action in or on the human body by pharmacological, immunological or metabolic means, but which may be assisted in its function by such means.

Minimum mandatory requirements: A list of the essential parameters and characteristics for which values or requirements have to be given in the detail specification.

Model lift: A representative lift whose technical dossier shows the way in which the essential safety requirements will be met for lifts which conform to the model lift defined by objective parameters and which uses identical safety components.

National Supervising Inspectorate (NSI): The authority that is responsible for completing an initial appraisal of inspection organisations, test laboratories, distributors and assessors, and the supervision of their operations subsequent to approval.

Nominal size (DN): A numerical designation of size which is common to all components in a piping system other than components indicated by outside diameters or by thread size. It is a convenient round number for reference purposes and is only loosely related to manufacturing dimensions. The nominal size is designated by DN followed by a number.

Non-automatic weighing instrument: A weighing machine that requires an operator during the weighing operation to obtain the result, e.g. to place the load on the load receptor of the instrument. By this definition, the following types of weighing instrument among others are included: retail shop scales, laboratory balances, bathroom scales, baby scales and kitchen scales.

Non-conformity: The non-fulfilment of specified requirements.

Non-standard item: An item which authorities have agreed not to make a standard item.

Normal use: An appliance is said to be 'normally used' when it is;

- correctly installed and regularly serviced in accordance with the manufacturer's instructions;
- used with a normal variation in the gas quality and a normal fluctuation in the supply pressure;
- used in accordance with its intended purpose or in a way which can be reasonably foreseen.

Notified Body: Notified Bodies are independent testing laboratories and/ or certification bodies recognised in the EU to perform tests and issue reports and certificates on conformity.

Ocean craft: Craft designed for extended voyages where conditions may exceed wind force 8 (Beaufort scale) and significant wave heights of 4 m and above, and vessels largely self-sufficient.

Offshore craft: Craft designed for offshore conditions up to and including wind force 8 and significant wave heights up to and including 4 m.

Operational cycle: A repeatable sequence of functional stresses.

Operational requirements: All the function and performance requirements of a product.

Organisation: A company, corporation, firm or enterprise, whether incorporated or not, public or private.

Personal protective equipment: Any device or appliance designed to be worn or held by an individual for protection against one or more health or safety hazards.

Phase: Defined segment of work.

Piping: Piping components intended for the transport of fluids, when connected together for integration into a pressure system. Piping includes in particular a pipe or system of pipes, tubing, fittings, expansion joints, hoses, or other pressure-bearing components as appropriate. Heat exchangers consisting of pipes for the purpose of cooling or heating air shall be considered as piping.

Placing on the market: The first making available in return for payment (or free of charge) of a device with a view to distribution and/or use on the EU market, regardless of whether it is new or fully furbished.

Placing on the market of the lift: When the installer first makes the lift available to the user.

Potentially explosive atmosphere: An atmosphere which could become explosive due to local and operational conditions.

Pre-inspections: This is an inspection for any obvious or physical damage such as broken meter glasses, knobs, bad dents to the case, broken fuse holders, disconnected wires etc. Where possible pre-inspections should be carried out on every equipment entering the workshop.

Pressure: Pressure relative to atmospheric pressure, i.e. gauge pressure. As a consequence, vacuum is designated by a negative value.

Pressure accessories: Devices with an operational function and having pressure-bearing housings.

Pressure equipment: Vessels, piping, safety accessories and pressure accessories. Where applicable, pressure equipment includes elements attached to pressurised parts, such as flanges, nozzles, couplings, supports, lifting lugs, etc.

Preventative maintenance: The maintenance that is carried out at predetermined intervals that are intended to reduce the probability of a failure occurring.

Procedure: Describes the way to perform an activity or process.

Product: Result of activities or processes (EN ISO 8402:1995).

Note 1 A product may include service, hardware, processed materials, software, or a combination thereof.

Note 2 A product can be tangible (e.g. assemblies or processed materials) or intangible (e.g. knowledge or concepts), or a combination thereof.

Note 3 A product can be either intended (e.g. offering to customers) or unintended (e.g. pollutant or unwanted effects.)

Product liability: A generic term used to describe the onus on a producer or others to make restitution for loss related to personal injury, property damage or other harm caused by a product or service.

Production lot: A quantity of components that have been manufactured continuously within a given period of time under uniform conditions.

Production permit: Written authorisation, prior to production or before provision of a service, to depart from specified requirements for a specified quantity or for a specified time.

Protective systems: Design units which are intended to halt incipient explosions immediately and/or to limit the effective range of explosion flames and explosion pressures. Protective systems may be integrated into equipment or separately placed on the market for use as autonomous systems.

Putting into service (as specified in Harmonised Directives): The stage at which a device is ready for use on the EU market for the first time for its intended purpose.

Qualification approval: The status given to a manufacturer's product unit, whose product has been shown to meet all the requirements of the product detail specification and Quality Plan.

Qualification approval certificate: A certificate that is issued to a component manufacturer that confirms qualification approval in respect of a specific electronic component or range of components.

Quality: The totality of features and characteristics of a product or service that bear upon its ability to satisfy stated or implied needs.

Quality assurance: All those planned and systematic actions necessary to provide adequate confidence that a product or service will satisfy given requirements for quality.

Quality assurance representative: The authorised representative of the National QA authority designated in the contract.

Quality audit: A systematic and independent examination to determine whether quality activities and related results comply with planned arrangements and whether these arrangements are implemented effectively and are suitable to achieve objectives.

Quality conformance inspection: Measures that are demanded in the specification to show that the components produced by a manufacturer fulfil the requirements of a particular specification.

Quality control: The operational techniques and activities that are used to fulfil requirements for quality.

Quality control system: The established management structure, responsibilities, methods and resources that together provide quality control to demonstrate the attainment of quality.

Quality level: A general indication of the extent of the product's departure from the ideal.

Quality loop: Conceptual model of interacting activities that influence the quality of a product or service in the various stages ranging from the identification of needs to the assessment of whether these needs have been satisfied.

Quality manager: A person who is responsible for the manufacturer's Quality Management System (also sometimes referred to as the Chief Inspector).

Quality management: That aspect of the overall management function that determines and implements the quality policy. **Note**: The terms 'quality management' and 'quality control' are considered to be a manufacturer/supplier (or first party) responsibility. 'Quality assurance' on the other hand has both internal and external aspects which in many instances can be shared between the manufacturer/supplier (first party), purchaser/customer (second party) and any regulatory/certification body (third party) that may be involved.

Quality Management System: The organisational structure, responsibilities, procedures, processes and resources for implementing quality management.

Quality Management System review: A formal evaluation by top management of the status and adequacy of the Quality Management System in relation to quality policy and new objectives resulting from changing circumstances.

Quality manual: A document setting out the general quality policies, procedures and practices of an organisation.

Quality plan: A document setting out the specific quality practices, resources and sequence of activities relevant to a particular product, service, contract or project.

Quality policy: The overall quality intentions and direction of an organisation as regards quality, as formally expressed by top management.

Quality procedure: A description of the method by which quality system activities are managed.

Quality programme: A documented set of activities, resources and procedures which implement the organisation's Quality Management System.

Quality records: Records should provide evidence of how well the Quality System has been implemented.

Quality spiral: Conceptual model of interacting activities that influence the quality of a product or service in the various stages ranging from the identification of needs to the assessment of whether these needs have been satisfied.

Quality surveillance: The continuing monitoring and verification of the status of procedures, methods, conditions, processes, products and services, and analysis of records in relation to stated references to ensure that specified requirements for quality are being met.

Quality verification inspections: These are performed prior to, during and after the job or task has been concluded. They are sometimes referred to as Pre-Inspections, In-Progress-Inspections and Out-Going Inspections.

Quarantine store: A secure place to store supplies that are awaiting proof that they comply with specified requirements.

Radio equipment: A product, or relevant component thereof, capable of communication by means of the emission and/or reception of radio waves utilising the spectrum allocated to terrestrial/space radiocommunication.

Radio waves: Electromagnetic waves of frequencies from 9 kHz to 3000 GHz, propagated in space without artificial guide.

Ratification: Formal acceptance of a NATO document (e.g. STANAG) as national implementing document.

Recreational craft: Means 'any boat of any type, regardless of the means of propulsion, from 2.5 to 24 m hull length, measured according to the appropriate harmonised standards intended for sports and leisure purposes'. The fact that the same boat could well be used for charter and/or for recreational boating training does not prevent it from being covered by this Directive when it is placed on the market for recreational purposes.

Redundancy: The existence, in a product, of more than one means of performing a function.

Refurbishing: The processing or reprocessing to specified requirements of a device, which has been previously released.

- the return of a device to the supplier;
- its modification by the supplier at the site of installation;
- its exchange; or
- its destruction;

in accordance with the instructions contained in an advisory notice.

Regression testing: Testing to determine that changes made in order to correct defects have not introduced additional defects.

Related documents: Documents referred to in a standard that form part of that standard.

Reliability: The ability of an item to perform a required function under stated conditions for a stated period of time.

Requirements of society: Requirements including laws, statutes, rules and regulations, codes, environmental considerations, health and safety factors, and conservation of energy and materials.

Safety: The freedom from unacceptable risks of personal harm.

Safety accessories (pressure equipment): Devices designed to protect pressure equipment against the allowable limits being exceeded. Such devices include:

- devices for direct pressure limitation, such as safety valves, bursting disc safety devices, buckling rods, controlled safety pressure relief systems (CSPRS), and
- limiting devices, which either activate the means for correction or provide for shutdown or shutdown and lockout, such as pressure switches or temperature switches or fluid level switches and 'safety related measurement control and regulation (SRMCR)' devices.

Safety component (also specified in Harmonised Directives): A component, provided that it is not interchangeable equipment, which the manufacturer or his authorised representative established in the EU places on the market to fulfil a safety function when in use and the failure or malfunctioning of which endangers the safety or health of exposed persons.

Sample: An item (or group of items) that have been taken from a larger collection (or population) of items to provide information relevant to that collection or population.

Sampling plan: An indication of the sample sizes and the acceptance/rejection criteria.

Sampling procedure: The operational requirements and instructions relating to the use of particular sampling plans or schemes.

Sampling scheme: The overall system containing a range of sampling plans and procedures.

Sampling size: The number of specimens in a sample.

Screening test: A test or combination of tests, intended to remove unsatisfactory items, or those likely to exhibit early failure.

Sealed pattern: A specimen electronic component that is taken from a lot which has successfully passed the qualification approval process and which is kept for subsequent reference.

Security: The prevention of use contrary to law and order.

Service liability: A generic term used to describe the onus on a producer or others to make restitution for loss related to personal injury, property damage or other harm caused by a product or service.

Sheltered water craft: Craft designed for small lakes, rivers, and canals where conditions up to, and including, wind force 4 and significant wave heights up to and including, 0.5 m.

Short-term: A device normally intended for continuous use for not more than 30 days.

Should: This auxiliary verb indicates that a certain course of action is preferred but not necessarily required.

Simple pressure vessel: Any welded vessel subjected to an internal gauge pressure greater than 0.5 bar which is intended to contain air or nitrogen and which is not intended to be fired.

Software: Covers all instructions and data which are input to a computer to cause it to function in any mode. This includes operating systems, supervisory systems, compilers and test routines, as well as application programmes. The words embrace the documents used to define and describe the programmes (including flow charts, network diagrams and programme listings), and also cover specifications, test plans, test data, test results and user instructions.

Specific: This adjective, when used with parameters or conditions, refers to a particular value or standardised arrangement, usually to those required in a European Standard or a legal requirement.

Specification: The document that describes the requirements with which the product, material or process has to conform.

Specified: This adjective, when used with parameters or conditions, refers to a particular value or standardised arrangement, usually to those required in a European Standard or a legal requirement.

Specified requirements: Either:

- requirements prescribed by the Purchaser and agreed by the Supplier in a contract for Product;
- requirements prescribed by the Supplier which are perceived as satisfying a market need;
- regulatory requirements.

Specimen: A representative item or quantity of material.

Standard: The result of a particular standardisation effort that has been approved by a recognised authority. When this word – in capital letters – appears in relation to a clause, letter or form, it means that the wording shall not be altered.

Standard item: An item which authorities agree should be used in preference to all others.

Standardisation: The process of formulating and applying rules for the benefit of all concerned.

Statistical quality control: That part of quality control in which statistical methods are used.

Statistical quality control chart: A method used to ensure that the performance of a product is maintained during manufacture whereby samples of the production (or process) are regularly analysed against a control chart that has the upper and lower permissible limits for that particular product or process, already plotted.

Storage life: The specified length of time prior to use for which items (which are known to be subject to deterioration) are deemed to remain fit for use.

Stress cycle: A repeatable sequence of stresses.

Structurally similar electronic component: Components which are made in one factory using virtually the same design, material, process and method of fabrication.

Supervising inspector: An inspector acting on behalf of the National Supervising Inspectorate.

Supplier: The organisation that provides a product to the customer (EN ISO 8402:1995).

Note 1 In a contractual situation, the supplier may be called the contractor.

Note 2 The supplier may be, for example, the producer, distributor, importer, assembler or service organisation.

Note 3 The supplier may be either external or internal to the organisation.

Note 4 With regard to MDD the term supplier is **not** used. The Directive instead refers to 'manufacturer'.

Supplier evaluation: Assessment of a supplier's capability to control quality.

Supplier rating: An index related to the performance of a supplier.

System review: The contractor's independent examination of the effectiveness of their system.

Technical construction file: A file describing the apparatus and providing information and explanations as to how the applicable essential requirements have been implemented.

Technician: An individual who is responsible for the actual maintenance, modification or repair of an item of equipment or product.

Telecommunications terminal equipment: A product enabling communication or a relevant component thereof which is intended to be connected directly or indirectly by any means whatsoever to interfaces of public telecommunications networks (that is to say, telecommunications networks used wholly or partly for the provision of publicly available telecommunications services).

Tender: Offer made by a supplier in response to an invitation to satisfy a contract to provide product.

Test: A critical trial or examination of one (or more) of the properties/characteristics of a material, product or service.

Test plan: A management document which addresses all aspects related to the test. It should include the test schedule and define the necessary support tools.

Test procedure: A document that describes each step that is necessary to conduct a test. The steps shall be in sequence with all the inputs and outputs defined.

Test specification: Describes the test criteria and the methods to be used in a specific test to assure that the performance and design specifications have been specified. The test specification identifies the capabilities or programme functions to be tested and identifies the test environment.

Toy: Any product or material designed or clearly intended for use in play by children of less than 14 years of age.

Traceability: The ability to trace the history, application or location of an item or activity, or similar items or activities, by means of recorded identification.

Transient: A device normally intended for continuous use for less than 60 minutes.

Type approval: The status given to a design that has been shown by type tests to meet all the requirements of the product specification and which is suitable for a specific application.

United Nations Recommendations: The recommendations laid down by the United Nations Committee of Experts on the Transport of Dangerous Goods, as published in the UN (*Orange Book*) and as amended by the date when this Directive is adopted; 'safety' shall mean the prevention of accidents and, where prevention fails, the containment of their effects.

User requirement: The documented product or service requirements of a customer or user.

Variable: A characteristic that is appraised in terms of values on a continuous scale.

Vendor appraisal: Assessment of a potential supplier's capability of controlling quality.

Verification: Confirmation by examination and provision of objective evidence that specified requirements have been fulfilled (EN ISO 8402:1995).

Note 1 In design and development, verification concerns the process of examining the result of a given activity to determine conformity with the stated requirement for that activity.

Note 2 The term 'verified' is used to designate the corresponding status.

Vessel: A housing designed and built to contain fluids under pressure including its direct attachments up to the coupling point connecting it to other equipment. A vessel may be composed of more than one chamber.

Volume (V): The internal volume of a chamber, including the volume of nozzles to the first connection or weld and excluding the volume of permanent internal parts.

Waiver: Written authorisation to use or release a quantity of material components or stores already produced but which do not conform to the specified requirements.

Weighing instrument: A measuring instrument that uses the action of gravity on a body to determine the mass of the body.

Work instruction: A description of how a specific task is carried out.

Workmanship: The level of the art or skill used in the repair process or manufacturing process as demonstrated by the characteristics of the product which cannot be specified in measurable terms.

Annex B: References

1 Standards

Number	*Date*	*Title*
ASC Q9000 series	various	Quality Management and Quality Assurance Standards
BS 0	1997	A Standard for Standards
BS 4778	1979	Quality Vocabulary
BS 4891	1972	A Guide to Quality Assurance
BS 5703:PT1	1980	Guide to Data Analysis and Quality Control Using Cusum Techniques – Introduction to Cusum Charting
BS 5750	1987	Superseded by ISO 9000:1994
BS 5760	various	Reliability of Systems, Equipment and Components
BS 6548 series		Maintainability of Equipment
BS 7000 series	various	Design Management Systems
BS 7750		Superseded by ISO 14001:1994
BS 7850 series	1992	Total Quality Management
EN 540	1993	Clinical Investigation of Medical Devices for Human Subjects
EN 724	1994	Guidance on the application of EN 29001 and EN 46001 and of EN 29002 and EN 46002 for non-active medical devices

Number	*Date*	*Title*
EN 928		In vitro diagnostic systems – guidance on the application of EN 29001 and EN 46001 and of EN 29002 and EN 46002 for in vitro diagnostic medical devices
EN 1041	1998	Information Supplied by the Manufacturer with Medical Devices
EN 29000	1987	Renumbered as ISO 9000/1
EN 46001	1996	Quality Systems – Medical Devices – particular requirements for the application of EN ISO 9001
EN 46002	1996	Quality Systems – Medical Devices – particular requirements for the application of EN ISO 9002
EN 50103	1995	Guidance on the application of EN 29001 and EN 46001 and of EN 29002 and EN 46002 for the active (including active implantable) medical device industry
IEC 271	1974	Guide on the reliability of electronic equipment and parts used therein – Terminology
ISO 3534 series	1977	Statistical terminology Part 1: Glossary of terms relating to probability and general terms relating to statistics.
ISO 8402	1994	Quality Management and quality assurance – Vocabulary
ISO 8800	No data	Health and Safety Management system recommendations
ISO 9000		Quality Management and Quality Assurance standards
ISO 9000/1	1994	Quality Management and Quality Assurance standards – guide to their selection and use
ISO 9000/2	1997	Quality Management and Quality Assurance standards – generic guidelines for the application of ISO 9001, 9002 and 9003

Number	*Date*	*Title*
ISO 9000/3	1997	Quality Management and Quality Assurance standards – guidelines for the application of ISO 9001 to the development, supply and maintenance of software
ISO 9000/4	1993	Quality Management and Quality Assurance standards – guide to dependability programme management
ISO 9001	1994	Quality Management Systems – model for Quality Assurance in design, development, production, installation and servicing
ISO 9002	1994	Quality Management Systems – model for Quality Assurance in production and installation
ISO 9003	1994	Quality Management Systems – model for Quality Assurance in final inspection and test
ISO 9004		Superseded by ISO 9004/1
ISO 9004/1	1994	Quality Management and Quality Management System elements – guide to quality management and Quality Management System elements
ISO 9004/2	1991	Quality Management and Quality Management System elements – guidelines for service
ISO 9004/3	1993	Quality Management and Quality Management System elements – guidelines for processed materials
ISO 9004/4	1994	Quality Management and Quality Management System elements – guidelines for quality improvement
ISO 10005	1995	Quality Management – guidelines for quality plans
ISO 10011/1	1990	Guidelines for auditing quality systems – auditing

Number	*Date*	*Title*
ISO 10011/2	1991	Guidelines for auditing quality systems – qualification criteria for quality systems auditors
ISO 10011/3	1991	Guidelines for auditing quality systems – management of audit programmes
ISO 10012/1	1992	Quality assurance requirements for measuring equipment – metrological confirmation system for measuring equipment
ISO 10012/2	1997	Quality assurance for measuring equipment – guidelines for control of measurement processes
ISO 10013	1995	Guidelines for developing quality manuals
ISO 11134	1994	Sterilisation of health care products – requirements for validation and routine control – industrial moist heat sterilisation
ISO 11135	1994	Medical devices – validation and routine control of ethylene oxide sterilisation
ISO 11137	1995	Sterilisation of health care products – requirements for validation and routine control
ISO 11737	1995	Sterilisation of Medical Devices
ISO 13485	1996	Quality systems – medical devices – particular requirements for the application of ISO 9001
ISO 13488	1996	Quality systems – medical devices – particular requirements for the application of ISO 9002
ISO 14001	1996	Environmental Management Systems – Specifications with guidance for use
ISO 14004	1996	Environmental Management Systems – General guidelines on principles, systems and supporting techniques

Number	*Date*	*Title*
ISO 14010	1996	Guidelines for Environmental Auditing – General Principles
ISO 14011	1996	Guidelines for Environmental Auditing – Auditing Procedures
ISO 14012	1996	Guidelines for Environmental Auditing – Qualification Criteria for Environmental Auditors

2 Other publications

Title	*Details*
A Positive Contribution to Better Business	British Standards Institution pamphlet on BS 5750/ISO 9000: 1987
An Executive Guide to the use of UK National Standards and International Standards for Quality Management Systems	A British Standards Institution publication
BSI in Europe	A British Standards Institution Newsletter
BSI Inspectorate	A British Standards Institution publication
BSI's Quality Management Handbook	(formally Handbook no 22 *Quality Management System*)
ISO 9000 for Small Businesses	Tricker, Ray. Published by Butterworth-Heinemann (1997)
Loss of quality through poor maintenance	A paper by HERNE Consultancy Group; (1994)
MDD Compliance Using Quality Management Techniques	Tricker, Ray. Published by Butterworth-Heinemann (1999)
Quality Assurance	A PSA publication printed in the UK for HMSO (1987), ISBN 86177.143.53

Title	*Details*
Selling to the Single Market	An HMSO Publication prepared for the Division of Trade and Industry and the Central Office of Information, June 1989
Standards in Electronics	Tricker, Ray: Butterworth-Heinemann (1996)
Statistical Process Control	Oakland, John S., Heinemann Newnes (1988)
Total Quality Control	Oakland, John S., (1989) Heinemann Newnes

Notes

Extracts from British Standards are reproduced with the permission of the British Standards Institute. Complete copies of all British Standards can be obtained, by post, from Customer Services, BSI Standards, 389 Chiswick High Road, London W4 4AL.

3 'New Approach' Website

A new website representing the joint efforts of CEN, CENELEC and ETSI, together with the European Commission and EFTA, has recently been developed, enabling 24-hour access to online information about European standards.

The 'New Approach' website provides SMEs with a facility for determining the appropriate standards for the products they manufacture, irrespective of which of the three European standards organisations is responsible for the applicable standards.

Additionally, all of the European standards referred to are available from BSI Customer Services on 0181 996 9001, while further information on particular products or markets may be obtained from BSI's Technical Help for Exporters advice service on 0181 996 7111. To access the New Approach website, go to www.NewApproach.org.

Annex C: Abbreviations and Acronyms

AFNOR	Association Francais de Normalisation
AG	General Assembly
AIMD	Active Implantable Medical Devices
ANSI	American National Standards Institute
AQL	Acceptable Quality Level
ASQC	American Society of Quality Control
ASTM	American Society Testing and Materials
BABT	The British Approvals Board for Telecommunications
BS	British Standards
BSI	British Standards Institution
BT	The Technical Board
BTAS/Q	Business and Technical Advisory Service on Quality
BTTF	Special Task Forces
BTWG	Working Groups
CA	Administrative Board
CB	Certification Body
CCA	Accord de Certification du CENELEC (CENELEC Certification Agreement International Radio Consultative Committee)
CCIR	International Radio Consultative Committee
CCITT	International Telegraph and Telephony Consultative Committee
CCQ	Comité de Coordination de Qualites
CE	Conformity Europe
CEB	Comité Electrotechnique Belge
CECC	CENELEC Electronic Components Committee (Comité des Composents Electroniques du Cenelec)
CEE	Commission Internationale de certification de conformité de l'equipment Electronique
CEI	Commission Electrotechnique Internationale
CEMEC	Committee of European Associations of Manufacturers of Electronic Components
CEN	Comité Europeen de Normalisation Electrotechnique – European Committee for Standardisation
CENELEC	European Committee for Electrotechnical Standardisation – combination of CENEL and CENELCOM (Comité Europeen de Normalisation Electrotechnique)

CEPT	European Conference of Postal and Telecommunications Administrators Certification Management Committee – of the IECQ
CMC	Canadian Standards Association
CR	CEN Reports
CTR	Common Technical Requirements
CWA	CEN Workshop agreements
DAR	German Accreditation Council
DIN	Deutsches Institut fur Normung e.v. (German Standards Institute)
DIS	Draft International Standard
DOA	Date of Announcement (EN Standards)
DOA	Dictionary of Abbreviations
DOP	Date of Publication (EN Standards)
DOR	Date of Ratification/Implementation (EN Standards)
DOW	Date of Withdrawal (EN Standards)
DTI	Department of Trade and Industry
E	Draft (Standard)
EC	European Community
ECAP	European Conformity Assessment Protocol
ECG	Electro-cardiogram
ECMA	European Computer Manufacturers Association
ECQR	Electronic Component Quality and Reliability Service
EDIG	European Defence Industry Group
EEA	European Economic Area
EEC	European Economic Community
EFTA	European Free Trade Association
EIA	Electronic Industries Association
EMAS	European Eco-Management and Audit Scheme
EMC	Electromagnetic Compatibility
EMI	Electromagnetic Interference
EMP	Electromagnetic Pulse
EMS	Environmental Management System
EN	European Normalisation
ENV	European Pre-standard
EOQC	European Organisation for Quality
EP	ETSI projects
EQD	Electrical Quality Assurance Directorate
EQFM	European Foundation of Quality Management
ER	Essential Requirement
ESCI	European Customer Satisfaction Index
ETSI	European Telecommunications Standards Institute
EU	European Union
EWOS	European Workshop for Open Systems
FMEA	Failure Mode and Effects Analysis
FTA	Fault Tree Analysis
HD	Harmonisation Document
HSC	Health & Safety Council
HSE	Health & Safety Executive

IEC	International Electrotechnical Commission
IEC	International Electrotechnical Commission – Quality Assessment System for Electronic Components
IECC	International Electrotechnical Commission Council
IECCA	Inter-Establishment Committee on Computer Applications
IECEE	International Electrotechnical Commission System for Conformity Testing
IECQ	Standards for Safety of Electrical Equipment
IEE	Institute of Electrical Engineers
IEEE	Institute of Electrical and Electronic Engineers
IEPG	Independent European Program Group
IIRS	Institute of Industrial Research and Standards (Irish NAI and NSI)
ILU	Integrated Logistic Unit
IMQ	Instituto Italiano del Marchio di Qualita (part of the Italian NSI)
IMS	Integrated Management System
ISO	International Organisation for Standardisation
IT	Information Technology
ITU	International Telecommunications Union
IVD	In Vitro Diagnostic
JIS	Japanese Industrial Standards
KEMAN-V	Kema (Netherlands NSI)
LVD	Low Voltage Directive
MDD	Medical Devices Directive
MRA	Mutual Recognition Agreement
MSD	Machine Safety Directive
NACCB	National Accreditation Council for Certification Bodies
NAF-AAD	National Authorisation Federation – Analogue and Digital
NAI	National Authorisation Institution
NAMAS	National Measurement Accreditation Service
NC	National Committee
NCB	National Certification Body
NCIQ	National Centre for Information Quality
NEC	Netherlands Elektrotechnishe Comité (Netherlands NAI)
NET	Normalisation de Europ en Telecommunications
NF	Norme Français
NQAA	National Quality Assurance Authority
NSA	National Supervising Authority
NSI	National Supervising Inspectorate
NSO	National Standards Organisation
OJ	See *OJEC*
OJEC	*Official Journal of the European Communities*
ONN	Organism National de Normalisation (another acronym for NSA)
ONS	Organism National de Surveillance (CECC acronym for NSI)
OSI	Open Systems Interconnection
PED	Pressure Equipment Directive
PPE	Personal Protective Equipment

PTN	Public Telecommunications Network
QA	Quality Assurance
QAMIS	Quality Assurance Management Information Systems
QASAR	Quality Management Systems Assessment and Registration
QC	Quality Control
QMS	Quality Management System
QP	Quality Procedure
QPL	Qualified Products List
RNE	French Accreditation Council
RTTE	Radio & Telecommunications Terminal Equipment
SAS	Swiss Association for Standardisation
SC	Subcommittee
SEM	Single European Market
SEV	Schweizerischer Elektrotechnischererein Swiss NAI and NSI)
SIE	Societe International d'electriciens
SIS	Swedish Institute for Standards
SNQ	Service National de la Qualité des Composants Electroniques (French NSI)
SPV	Simple Pressure Vessel
STF	Specialist Task Force
TC	Technical Committee
TCF	Technical Construction File
TQC	Total Quality Control
TQM	Total Quality Management
TTE	Telecommunications Terminal Equipment (Directive)
UK	United Kingdom
USA	United States of America
USE	Union des Syndicats de l'Electricité
UTE	Union Technique de l'Electricité (French NAI)
VD	Verband Deutscher Elektrotechniker (German NSI)
WAMH	Workplace Applied Modular Harmonisation
WAUILF	Workplace Applied Uniform Indicated Low Frequency (Application)
WELMEC	The Western European Legal Metrology Co-operation
WI	Working Instruction
YFR	Yearly Functional Rules
YFL	Yearly Functional Logistics

Annex D: Addresses

1 Availability of published standards

1.1 European Standards Organisations

CEN

36, Rue de Stassart
B – 1050 Brussels
Tp: +32 (2) 550.08.11
Fx: +32 (2) 550.08.19
E-mail: **infodesk@cenclcbel.be**
URL address:
http://www.cenorm.be/

CENELEC

35, Rue de Stassart
B – 1050 Bruxelles
Tp: +32 (2) 519.68.71
Fx: +32 (2) 519.69.19
E-mail: **general@cenelec.be**
URL address:
http://www.cenelec.be/

ETSI

Route des Lucioles – Sophia Antipolis – Valbonne
F – 06921 Sophia Antipolis Cedex
Tp: +33 (0)4 92 94 42 00
Fx: +33 (0)4 93 65 47 16
E-mail: **infocentre@etsi.fr**
URL address: **http://www.etsi.org/**

1.2 Government and Non-Government Organisations

American National Standards Institute

11 West 42nd Street, 13th Floor
New York, NY 10036
(212) 642–4900
Fax (212) 398–0023

American Society for Quality Control	611 East Wisconsin Avenue PO Box 3005 Milwaukee WI 53201–3005 (414) 272–8575 or (800) 248–1946 Fax (414) 272–1734
National Center for Standards and Certification Information	US Department of Commerce Building 820, Room 164 Gaithersburg, MD 20899 (301) 975–4040 EU Hotline (301) 921–4164 Fax (301) 926–1559
European Organisation for Testing and Certification (EOTC)	Egmont House Rue d'Egmont 15 1000 Brussels, Belgium (32 2) 502–4141 Fax (32 2) 502–4239
International Electrotechnical Commission (IEC)	Rue de Varembe 3 Case Postale 131 1211 Geneva 20, Switzerland (41 22) 919–0211 Fax (41 22) 919–0300
International Standards Organization (ISO)	Rue de Varembe 1 Case Postale 56 1211 Geneva 20, Switzerland (41 22) 749–0111 Fax (41 22) 733–3430

1.3 Commercial Sales Agents

British American Chamber of Commerce	41 Sutter Street, Suite 303 San Francisco, CA 94104 (415) 296–8645 Fax (415) 296–9649
Compliance Engineering	One Tech Drive Andover, MA 01810 (508) 681–6600 Fax (508) 681–6637
Custom Standards Services, Inc.	310 Miller Avenue Ann Arbor, MI 48103 (734) 930–9277 or (800) 699–9277 Fax (734) 930–9088

European Document Research	1100 17th Street NW, Suite 301 Washington, DC 20036 (202) 785–8594 Fax (202) 785–8589
Global Engineering	(A subsidiary of IHS, listed below) 7730 Carondelet Avenue, Suite 407 Clayton, MO 63105 (314) 726–0444 or (800) 854–7179 Fax (314) 726–6418
Information Handling Services (IHS)	15 Inverness Way East Englewood, CO 80112 (303) 790–0600 or (800) 525–7052 Fax (303) 397–2599
NMi USA, Inc.	36 Gilbert Street South Tinton Falls, NJ 07724 (732) 842–8900 Fax (732) 842–0304
QNET LLC	(CE Marking, Training & Consulting) P. O. Box 527 Elk River, MN 55330–0527 Phone: (612) 441–0899 Fax: (612) 441–0898 E-mail: **qnet@ce-mark.com**
Qualified Specialists, Inc.	363 North Belt, Suite 630 Houston, TX 77060 (281) 448–5622 Fax (281) 448–6015
Simcom International	6111 Peachtree Dunwoody Road Building E., Suite 200 Atlanta, GA 30328–4577 (770) 730–9980 Fax (770) 730–9976 E-mail: **simcom@cemark.com**
SWBC America, Inc.	4938 Hampden Lane, #226 Bethesda, MD 20814 (301) 656–9125 Fax (301) 656–9128 E-mail: **SWBCusa@aol.com**

2 Availability of national standards

The text of European Standards, transposed as national standards, can be obtained from the following National Standards Bodies:

Österreichisches
Elektrotechnisches Komitee (ÖEK)
Österreichischen Verband für
Elektrotechnik (ÖVE)
Eschenbachgasse 9
A – 1010 Vienna
http://www.ove.at
Tel: +43 1 587 63 73
Fax: +43 1 586 74 08
E-mail: **ove@ove.at**

Comité Electrotechnique Belge
(CEB)
Belgisch Elektrotechnisch Comité
(BEC)
avenue Fr. Van Kalken 9
B – 1070 Bruxelles
http://www.bec-ceb.be
Tel: +32 2 556 01 10
Fax: +32 2 556 01 20
E-mail:
centraloffice@bec-ceb.be

Czech Standards Institute (CSNI)
Biskupsky dvûr 5
CZ – 110 02 Praha 1
http://www.csni.cz
Tel: +420 2 21 80 21 00
Fax: +420 2 21 80 23 11
E-mail: **csni@login.cz**

Dansk Standard (DS)
Electrotechnical Sector
Kollegievej 6
DK – 2920 Charlottenlund
http://www.ds.dk
Tel: +45 39 96 61 01
Fax: +45 39 96 61 02
Fax: +45 39 96 61 03
(Certification dept.)
E-mail: **standard@ds.dk**

Finnish Electrotechnical
Standards Association (SESKO)
Särkiniementie 3
P.O. Box 134
SF – 00211 Helsinki
http://www.sesko.fi
Tel: +358 9 696 391
Fax: +358 9 677 059
E-mail: **finc@sesko.fi**

Union Technique de l'Electricité
(UTE)
33, Av. Général Leclerc – BP 23
F – 92262 Fontenay-aux-Roses
Cedex
http://www.ute-fr.com
Tel: +33 1 40 93 62 00
Fax: +33 1 40 93 44 08
E-mail: **ute@ute.asso.fr**

Deutsche Elektrotechnische
Kommission im DIN und VDE
(DKE)
Stresemannallee 15
D – 60 596 Frankfurt am Main
http://www.dke.de
Tel: +49 69 63 080
Fax: +49 69 63 12 925
E-mail: **dke.zbt@t-online.de**

Hellenic Organization for
Standardization (ELOT)
Acharnon Street 313
GR – 111 45 Athens
http://www.elot.gr
Tel: +30 1 212 01 00
Fax: +30 1 228 30 34
E-mail: **elotinfo@elot.gr**

The Icelandic Council for Standardization (STRI)
Holtagardar
IS – 104 Reykjavik
http://www.stri.is
Tel: +354 520 71 50
Fax: +354 520 71 71
E-mail: **stri@stri.is**

Electro-Technical Council of Ireland (ETCI)
Unit 43
Parkwest Business Park
IRL – Dublin 12
http://www.etci.ie
Tel: +353 1 623 99 01
Fax: +353 1 623 99 03
E-mail: **administrator@etci.ie**

Comitato Elettrotecnico Italiano (CEI)
Viale Monza 259
I – 20126 Milano
http://www.ceiuni.it
Tel: +39 02 25 77 31
Fax: +39 02 25 77 32 10
E-mail: **cei@ceiuni.it**

Service de l'Energie de l'Etat (SEE)
B.P. 10
L – 2010 Luxembourg
http://www.etat.lu/SEE
Tel: +352 46 97 461
Fax: +352 46 97 46 – 39
E-mail:
see.normalisation@eg.etat.lu

Nederlands Elektrotechnisch Comité (NEC)
Kalfjeslaan 2
Postbus 5059
NL – 2600 GB Delft
http://www.nni.nl
Tel: +31 15 269 03 90
Fax: +31 15 269 01 90
E-mail: **info@nni.nl**

Norsk Elektroteknisk Komite (NEK)
Harbitzalléen 2A
Postboks 280 Skoyen
N – 0212 Oslo
http://www.nek.no
Tel: +47 22 52 69 50
Fax: +47 22 52 69 61
E-mail: **nek@nek.no**

Instituto Português da Qualidade (IPQ)
Rua Antorio Ciao 2
P – 2829–513 Caparica
http://www.ipq.pt
Tel: +351 1 294 81 00
Fax: +351 1 294 81 81
E-mail: **ipq@mail.ipq.pt**

Asociación Española de Normalización y Certificación (AENOR)
C/Génova 6
E – 28004 Madrid
http://www.aenor.es
Tel: +34 1 432 60 00–432 60 23 (Info Service)
Fax: +34 1 310 45 96–310 36 95 (Standt Dept)
E-mail: **norm.clciec@aenor.es**

Svenska Elektriska
Kommissionen (SEK)
Kistagången 19
Box 1284
S – 164 28 Kista Stockholm
http://www.sekom.se
Tel: +46 84 44 14 00
Fax: +46 84 44 14 30
E-mail: **snc@sekom.se**

Swiss Electrotechnical Committee
(CES)
Luppmenstraße 1
CH – 8320 Fehraltorf
http://www.sev.ch
Tel: +41 1 956 11 11
Fax: +41 1 956 11 22
E-mail: **sev@sev.ch**

British Electrotechnical
Committee (BEC)
British Standards
Institution (BSI)
389 Chiswick High Road
GB – London W4 4AL
http://www.bsi.org.uk
Tel: +44 181 996 90 00
Fax: +44 181 996 74 60

3 Information sources

American National Standards
Institute (ANSI)
11 West 42nd Street
New York, NY 10036
Tel: 212–642–4900
Fax: 212–302–1286

British Standards Institute (BSI)
PO Box 16206
Chiswick
London, W4 4ZL, UK
Tel: 011–44–181–996–7000
Fax: 011–44–181–996–7001

ILI
Index House
Ascot
Berkshire
SL5 7EU, UK
Tel: 011–44–1344 636400
Fax: 011–44–1344 291194

Office of the European Union and
Regulatory Affairs
US Dept of Commerce –
Room 3036
14th and Constitution Avenue,
N.W.
Washington, D.C. 20230
Tel: 202–482–5276
Fax: 202–482–2155

International Electrotechnical
Commission (IEC)
3, Rue de Varembe,
P.O. Box 131
1211 Geneva 20, Switzerland
Tel: 011–41–22–919–0228
Fax: 011–41–22–919–0300

International Organisation for
Standardisation (ISO)
Case Postal 56
1222 Geneva 20, Switzerland
Tel: 011–41–22–7490336
Fax: 011–41–22–7341079

VDE-Verlag GmbH
Bismarkstrasse 33
10625 Berlin, Germany
Tel: 011–49–30–348001–0
Fax: 011–49–30–3417093

Office of the Official Publications
of the EC
2 Rue Mercier
L2144 Luxembourg
Tel: 011–352–29291
Fax: 011–352–292942763

StingRay Management
Consultants
Riddiford House
Winkleigh
Devon
EX19 8DW
Tel: (+44) 1837 93118
Fax: (+44) 1837 83118

About the Author

Ray Tricker (MSc, IEng, FIIE(elec), FinstM, MIQA, MIRSE) as well as being the Principal Consultant and Managing Director of Herne European Consultancy Ltd – a company specialising in ISO 9000 and ISO 14000 Management Systems – is also an established Butterworth-Heinemann author. He served with the Royal Corps of Signals (for a total of 37 years) during which time he held various managerial posts culminating in being appointed as the Chief Engineer of NATO ACE COMSEC.

Most of Ray's work since joining Herne has centred on the European Railways. He has held a number of posts with the Union International des Chemins de fer (UIC) (e.g. Quality Manager of the European Train Control System (ETCS), European Union (EU) T500 Review Team Leader, European Rail Traffic Management System (ERTMS) Users Group Project Co-ordinator, HEROE Project Co-ordinator) and is currently preparing a complete Quality Management System for the European Rail Research Institute (ERRI) in Holland, aimed at gaining them ISO 9000 accreditation in the near future. He is also consultant to the Association of American Railroads (AAR) advising them on ISO 9001:2000 compliance.

Index